INTRODUCTION TO PETROLEUM PROCESS SAFETY

CHIDI VENANTIUS EFOBI

PARTRIDGE

ISBN:	Hardcover	978-1-5437-5933-4
	Softcover	978-1-5437-5932-7
	eBook	978-1-5437-5934-1

Print information available on the last page.

To order additional copies of this book, contact
Toll Free +65 3165 7531 (Singapore)
Toll Free +60 3 3099 4412 (Malaysia)
orders.singapore@partridgepublishing.com

www.partridgepublishing.com/singapore

DEDICATED TO

God Almighty, for His grace and favors;

Ezioma, my amiable wife, for her love, affection, understanding, patience and encouragement;

Chimazuru (Chima) and Chizitere (Chizi), our children, for the affection, warmth and joy they bring;

Ferdinand, my late Dad, for being a real dad and teaching me the basic art of writing;

Celine, my late mum and first grade teacher, for being a real mum and teaching me to be disciplined;

Uncle Arthur Aso, for inspiring me to be an engineer.

CONTENTS

FOREWORD

There is no doubt that the discovery of petroleum has provided a significant impetus to the progression of the industrial revolution. However, despite unprecedented advances in technology and safety practices, catastrophic incidents continue to blight the process industry.

The demand is ever increasing to not only learn from incidents but also to build a proactive approach that is capable to predict an incident and ultimately prevent it from happening. This philosophy is the pinnacle of process safety and comprised of layers of safety principles and engineering solutions that work together to constitute a robust and practical safety system. Such a comprehensive safety management system is the foundation to achieve process safety excellence resulting in profitability and delivering sustainable long-term value to stakeholders.

It is essential that young and emerging petroleum engineers are introduced to the concept of process safety and this book provides an introduction to process safety principle in the hydrocarbon industry. Although some of the major process safety incidents are discussed at the beginning of the book, the reader will be introduced to the staples of process safety in a structured manner to emphasize on the importance of being proactive rather than reactive. This book is a valuable resource in the field of process safety and is targeted to introduce the elements of process safety to safety engineering students as well as loss prevention engineers within the petroleum industry.

Ghassan G. Abulfaraj
Manager of Loss Prevention
Saudi Aramco

CHAPTER 1

INTRODUCTION

Chemicals have been made and used one way or the other from ancient times. However, the production and use of chemicals in large quantities started around the period of the industrial revolution in the early to mid-19th century. Around the same period the first commercial oil well was drilled at a site on Oil Creek near Titusville, Pennsylvania. These developments were good but they have also resulted in major accidents with catastrophic consequences to workers and the public, assets and the environment. These kinds of accidents need to be prevented or the consequences minimized. Process safety covers how major hazards arising from process industries are identified, assessed and controlled. It is a blend of engineering and management systems that is geared towards preventing (or minimizing the consequences of) loss of containment of hazardous substances, fires, explosions, structural collapses and other catastrophic accidents in the process industries (like chemicals and petroleum).

Below are some of the worst process industry accidents in recent history.

1. The Flixborough Disaster

Nypro UK (a joint venture between Dutch State Mines and the British National Coal Board) owned a chemical plant close to the village of Flixborough, England. It was originally designed to produce fertilizer from by-products of the coke ovens in a nearby steel mill. It

1

was later retrofitted to produce caprolactam, a type of chemical used to manufacture nylon 6. As part of the process, cyclohexane was heated and compressed air infused to oxidize into cyclohexanone (and some cyclohexanol). This process took place in a series of six reactors.

On March 27 1974, a crack was discovered in reactor No. 5 and cyclohexane was leaking from it. The facility was shut down and the situation assessed. It was consequently decided to bypass the reactor in order to allow operations to continue while arrangements are made to repair the leak. The bypass line linking reactor No. 4 to No. 5 was fabricated out of a 20-inch nominal bore pipe (DN 500 mm). It was pressure-tested at the working pressure using nitrogen as the test medium and subsequently commissioned.

After two months of satisfactory operation, the reactors were shut down, depressurized and cooled in order to mend some other leaks in another part of the plant. On June 1 1974, after the new leak had been mended, the process of restarting the plant commenced. In the afternoon, the 20 inch bypass system ruptured, huge amount of hot cyclohexane was released and ignited, leading to massive explosion. The impact blew off the roof of the control building, shattered the windows and killed all the 18 workers inside. There were a total of 28 fatalities and 36 serious injuries inside the plant. Luckily no fatalities were reported outside the plant but there were 50 injuries and over 2000 properties destroyed. The inferno raged for several days before it was put under control.

Some of the key technical failures that caused this disaster (and the disastrous consequences) were as follows:

- Inadequate management of change process – plant was modified without adequately assessing the potential consequences;
- Sub-standard design – the design engineers did not consider the potential for a major disaster in the layout of the facilities and the structural design of the control building;
- Less than adequate startup procedure – this incident happened during startup of the plant. The startup procedure was found to be less than adequate.

2. Mexico City LPG Tank Farm Explosion

This incident happened at a liquefied petroleum gas (LPG) tank farm in San Juanico, near Mexico City, Mexico. This facility which belonged to Petroleos Mexicanos (PEMEX) consisted of 54 LPG storage tanks; 6 large spherical tanks and 48 smaller horizontal bullet shaped tanks, the total storage capacity being 11,000 m^3.

The incident happened on November 19 1984. Early in the morning of the fateful day there was gas leakage from a pipe rupture due to excessive pressure. A gathering vapor cloud was moved by the wind towards the ground flare pit located at the corner of the plant. At about 5:40 hours the vapor cloud was ignited leading to an extensive conflagration in the plant. About 4 minutes later there was the first explosion followed by over a dozen others within the first hour. Some of these explosions were of the BLEVE type (Boiling Liquid Expanding Vapor Explosion) when the storage tanks ruptured. The impact of these explosions damaged many houses, sent metal fragments flying distances of up to 1200 m from the plant and further fueling the fire with more gas. It took over 24 hours for the fire to come under control. The plant was completely destroyed, 500 to 600 people were killed and 5,000 to 7,000 people severely injured.

3. Bhopal Disaster

Union Carbide Corporation (UCC) built a plant for the manufacture of pesticides in Bhopal, India in 1970. The local subsidiary Union Carbide India Limited (that owned the plant) has UCC and Indian Government as key shareholders. The plant location was not zoned for hazardous industries but for light industrial and commercial use. The shareholders influenced the approval of the site because of its central location and easy access to transport infrastructure. In addition, the plant was approved for the manufacture of pesticides from component chemicals imported from the parent company overseas. One of such chemicals is known as methyl isocyanate (MIC). However, due to stiff competition and unfavorable business environment, the local company decided to incorporate the processing of the component chemicals (which involves more complex and hazardous processes) in the Bhopal facility. That notwithstanding, the fortunes of the local company did not improve. Due to financial pressure the integrity of several safety equipment were compromised and safety procedures were also bypassed. The local regulatory authorities were aware of these but decided to look the other way for fear of pushing the company out of business with attendant job losses.

Close to midnight on December 2, 1984, an operator observed a small leak of MIC gas and a steady rise in pressure inside a storage tank. A safety device (vent –gas scrubber) designed to neutralize toxic releases from the MIC system was turned off some weeks earlier. Some water for internal pipes cleaning passed through a faulty valve and mixed with the MIC. The resultant reaction was exothermic with rapid increase in pressure and temperature. A refrigeration unit that was designed to cool the MIC storage system had been dismantled and relocated to another section of the plant. The gas flare system was also not functional. At about one hour after midnight, a safety valve gave way and a plume of MIC gas was released into the atmosphere, drifted into the local community and killed over 3,800 people and hundreds of domestic animals immediately. In the years that followed it was estimated that further 20,000 people died and scores left with health problems traceable to the incident.

Due to mediation by the India Supreme Court and the government, UCC paid compensation of $470 million to the victims and survivors of the deceased. The plant is closed till this day.

4. Piper Alpha

Piper Alpha, a major North Sea oil and gas drilling and production platform was built in 1976. It was initially built as an oil platform with capacity of about 250,000 barrels of oil per day increasing to 300,000 barrels and eventually declining to 125,000 barrels by 1988. In 1980, a gas recovery module was installed on the platform. The operating company (Occidental) built an oil terminal (Flotta) in the Orkney Islands to receive and process oil from Piper Alpha and two other platforms (Claymore and Tartan). A 30-inch diameter main oil pipeline carried the oil from Piper Alpha to the terminal, with a line from Claymore joining it downstream. Tartan platform fed oil to Claymore and also onto the main line to Flotta. A separate 18-inch diameter gas pipeline went from Piper Alpha to the Tartan and from Piper to a gas compression platform (known as MCP-01). It is noteworthy that on installation, Piper Alpha was built meeting the standards of the day. For example, the modules were separated such that the most dangerous operations were distant from where personnel stayed. When the gas recovery module was introduced, the change was not properly managed to ensure that safety standards were still being maintained. The gas compression module was sited next to the control room.

On 6th of July 1988 a series of events started that would end up destroying the platform and killing 165 workers on board and 2 crew members from a rescue craft that was responding to the emergency. On the fateful day, drilling, production, inspection and maintenance (some by divers) were on going. One of the platform's two condensate pumps was out of service for maintenance work on a pressure relief valve. The work on the pressure relief valve could not be completed by the shift crew that started it and a decision was taken to continue the next day. The condensate pipe was then sealed with a blind flange but the status was not communicated to the incoming shift. The work was being done under the work permit system. In the night (when the next shift had taken over), the second pump tripped and the flare intensity increased. In order to avoid shutting down the facility, the first pump was started up. The crew did not know there was a missing pressure relief valve and a blind flange was in its place. The blind flange leaked, gas ignited and caused an explosion in the gas compression module. This explosion wiped out the control room because the separation distance from the compression module was not adequate and the control room had no blast walls. Large oil fires from the topside inventory was initiated causing intense smoke and fires. Meanwhile, the fire water system could not start because it had been put on manual mode because of the divers in the water. Two men that risked trying to go and start the pumps never made it alive. As the fire continued and smoke intensified, the personnel on board started gathering in the accommodation area but there was no clear direction on what to do. Panic ensued and some decided to use their own initiative to jump into the water.

Meanwhile the Tartan platform continued pumping gas to Piper Alpha because the personnel on board (Tartan) had no authority to shut down. The intense heat from the inferno melted the riser and it ruptured, causing a huge fireball that completely engulfed the platform. At this time, the helideck was inaccessible to helicopters and the lifeboats could not get close as well. Hell had broken loose and Piper Alpha was completely consumed. Only 61 of the 226 personnel on board, who took their fate into their hands survived the distance. Property damage was estimated to be about 3.4 billion US Dollars.

Investigation of the accident revealed the following key failures:

- Work permit was not properly used;
- Piper Alpha and surrounding platform installation managers lacked authority to shut down the facilities;
- Incident command system was not robust enough and failed when put to test;
- Inadequate shift turnover;
- Fire water system was on manual mode and no proper way of starting it in an emergency;
- Lack of (or inadequate) change management process, as the associated risks of adding the gas recovery was not properly analyzed;
- Separation distance of the control room from hazardous operations was inadequate and the room was unprotected;
- No segregation of hazardous areas from non-hazardous;
- Risers were inappropriately sited and not fire-proofed;
- The refuge area and escape systems were inadequate.

5. Phillips Pasadena Disaster

This incident occurred in a Phillips chemical complex located at Pasadena, near Houston. The facility produced high-density polyethylene (HDPE) and employed 905 people. In addition to the regular employees, there were about 600 daily contract personnel that carried out regular maintenance activities and construction work in the plant expansion project.

The incident happened during routine maintenance activities on October 23, 1989 on one of the polyethylene reactors. As part of the isolation process, the air connections for opening and closing the isolation valves were supposed to be disconnected. However, on the reactor, the air connections had been improperly interconnected. Consequently, the valve that was seen as closed in the control room was actually open, releasing highly flammable process gases into the atmosphere.

Within a short period, a vapor cloud had formed, moving quickly through the plant. In the process, the vapor cloud was ignited and an explosion followed. Several other explosions and massive fire did extensive damage to the plant, killed 23 people and injured 314. The inferno took over 10 hours to be brought under control. The cost of the facilities destroyed was estimated at over $715 million and loss due to business disruption, $700 million.

6. Texas City Oil Refinery Explosion

Texas City Oil Refinery was built in 1934 and initially owned by Amoco. It was however acquired by BP in 1999 when it bought over Amoco. The latter had a cost saving management style and hence turnaround maintenance works, scheduled inspections, and major upgrades were usually allowed to slip. Early in 2005 an independent consultant had assessed the plant as accident waiting to happen as the integrity of several safety critical elements were in question. Consequently, several turn around maintenance jobs were started in several units of the refinery.

On the date of the incident (March 23 2005), the turnaround maintenance works in the isomerization unit had been completed and start-up process had commenced. At about 2:00 am (during the night shift) operations personnel opened the valve to start the release of highly flammable liquid hydro-carbons into a raffinate splitter tower. Due to faulty level transmitters, opaque sight glass and dysfunctional safety alarm the tower was overfilled to over 20 times the normal level. At about 09:30 am the day shift personnel took over without proper turnover. The new shift personnel continued to pump hydrocarbon into the vessel and at about 12:40 pm a high pressure alarm was triggered.

Operators intervened and opened the tower's outtake valve (rather very late). Prior to this time, the furnaces had been turned on to pre-heat raffinate going into the tower and to heat the raffinate at the bottom of the vessel. The procedure for this heating was not followed and hence the liquid was heated far beyond specifications. On opening the outtake valve, the liquid was quickly released into the heat exchanger. The overheating and quick release overloaded the heat exchanger causing the liquid going into the tower to be much hotter than normal. When this liquid mixed with the already high level in the tower, it boiled over, expanded, spilled out of the safe-guard release tube and via the blow-down stack into the atmosphere. Within minutes a large flammable vapor cloud had developed and at about 1:20 pm, sparks from a running diesel truck nearby ignited it resulting into series of explosions. Fifteen workers were killed and one hundred seventy injured. BP paid $50 million in fines.

Investigation into the accident concluded that there were organizational and technical failures. The key organizational failures included undue cost-cutting, lack of adequate investment in plant infrastructure, lack of corporate oversight on major accident prevention programs, undue emphasis on occupational safety (with less emphasis on process safety), defective management of change process, lack of competent supervision for startup operations, inadequate training of operators, poor communications, and non-adherence to approved procedures. The key technical failures were as follows: undersized blowdown drum, inadequate preventive maintenance of safety critical systems, non-calibration of level sensors, inoperable alarms, and use of outdated blowdown drum and stack technology.

7. Bouncefield Tank Farm Explosion

Bouncefield oil tank farm is a petroleum products storage and transfer facility in Hemel Hempstead, Hertfordshire, England.

At 18:50 hours on December 10 2005, delivery of gasoline into one of the tanks that has a capacity of 6 million liters started. The tank had two level controls. One of them was an automatic tank gauging system (ATG) that measured and transmitted the level to a control panel. The second control was an independent high-level switch (IHLS) that has the capability of sounding audible alarms and closing the valve on the delivery line when the level got to the set-point.

At about 03:05 hours on December 11 2005, the ATG got stuck in position. The implication of this was that even though the level of liquid in the tank was rising the change could no longer be communicated to the control panel. The three ATG alarms (user, high and high-high levels alarms) could also not work. So the gasoline levels continued to rise without any action. Investigation later revealed that the ATG had stuck in position several times in the past but the deficiency was not effectively addressed. The IHLS could also not operate as required, as it was not properly installed. Those that installed the switch missed out a critical component (a padlock that was required to retain a check lever in a working position) as they did not fully understand how it worked. By 05:37 hours on December 11 2005, the tank got filled-up and gasoline started spilling out from the vents at the tank roof. Over 250,000 liters spilled into the bund wall unnoticed. Vapor cloud that developed was observed by members of the public who alerted the employees. The fire alarm was sounded and firewater pumps started at 06:01 hours. Almost immediately, the vapor cloud was ignited and exploded. It took several days to extinguish the fire.

There was no fatality but over 40 were wounded. Twenty tanks in the farm were completely burnt and third party property in the local community severely damaged by the explosion leading to evacuation of homes and businesses. The bund wall around the tank and the drains system that were designed to control the release of spills failed, leading to further environmental pollution. The companies involved: Hertfordshire Oil Storage Limited, British Pipeline Agency, Motherwell Control Systems, TAV Engineering & Total UK Limited paid varying fines totaling over $1,400 million.

8. Deepwater Horizon Blowout

The Deepwater Horizon was a semi-submersible mobile floating rig owned by Transocean and had the capacity to operate in waters up to 3,000 meters (10,000 feet) deep. It was hired by BP to work from

March 2008 to September 2013. It was drilling an exploratory well 66 kilometers (41 miles) off the coast of Louisiana, United States of America (in the Gulf of Mexico) at a water depth of approximately 1,600 meters (5,100 feet).

At about 9:45 pm (local time) on April 20 2010, high pressure methane gas from the formation expanded into the wellbore, rose into the drilling rig, ignited and exploded. The resulting inferno quickly engulfed the entire platform. Majority of the 126 crew members on board the rig were rescued by helicopters or lifeboats. Eleven workers are thought to have been consumed in the explosion as their bodies were not found despite serious search operations. The rig sank after two days.

After the explosion and sinking of the rig, oil gushed out from the well bore for 87 days, discharging an estimated 4.9 million barrels of oil into the Gulf of Mexico until it was capped and sealed. Aside from personnel injuries and multiple fatalities, the accident caused massive damage to the environment and reputation of BP. The company chief executive officer lost his job as a result. Some people in the affected areas lost their means of livelihood. It was considered to be the largest marine oil spill in the history of the petroleum industry.

Investigation into the cause of the gas leak revealed defective cement on the well. Root causes were traceable to series of cost-cutting decisions that culminated in inadequate safety systems. BP paid $18.7 billion in fines and over $40 billion in criminal and civil settlements.

CHAPTER 2

PETROLEUM FACILITIES AND PROCESSES

Reservoir fluids from producing wells are transported via flow-lines to processing facilities. The fluid consists of crude oil, gas, water and various contaminants (like salt and hydrogen sulfide). The percentages of these constituents vary from place to place. Petroleum production facilities are designed to separate/process crude oil and gas to meet customers' specifications. The facilities and processes described here do not apply in every location. The designs vary depending on the properties of the reservoir fluid, geographical location, local laws, economic/ financial and other consideration. However, a generally accepted rule in the designs is to process petroleum as close as possible to the field so as to ease transportation and minimize damage to transport facilities (e.g. pipelines) by various contaminants.

Separation Facilities

The first stage in petroleum processing is separation into crude oil, gas and water. This is normally done using horizontal pressure vessels known as separators. Flow-lines carrying reservoir fluid from each well meet at an inlet manifold or production header. With the use of valves, each well is routed into one or more of several manifold headers. Separation plants have at least one production manifold header and a test header. Plants with more than one train have more than one production header. The test header is used to route one or more wells to the test separator. Aside from the well undergoing test at any point in time, the rest are routed to a first stage separator or production trap.

Fluids come in from the wells at high pressure and the separators are set at lower pressures. Inside the separator vessels are also installed baffles and demister pad. On entering the vessel, the drop in pressure aided further by baffles and demister pads separate the three fluid phases. Water settles at the bottom of the separator, oil floats on top and gas is flashed out from the top. It is noteworthy that this separation is typically in stages because a huge pressure reduction in a single separator may cause flash vaporization, with consequence of instability and attendant safety hazards. Separated oil from the first stage separator may be cooled or heated in a heat exchanger and passed through a lower pressure second stage separator to further separate oil, gas and water. This may be repeated in several stages, successive stages operating at lower pressures than the previous. The more the stages the less the dissolved gas and lower the flash point of the final oil product in-order to meet customers' specifications. The gas flashed out is sent either to be flared or to the compression stage.

A test separator is used to separate fluid from one or more wells designated for analysis and detailed flow measurement. This test is conducted when a well is first taken into production and subsequently at regular intervals (normally 1-2 months). The test measures the total and component flow rates under different production conditions. Samples from each well are also taken and analyzed in the laboratory to determine the composition.

Dehydration Facilities

Wet crude oil from the last stage separator is pumped to the dehydration equipment to further remove entrapped water. There are various types of dehydration equipment, some of which are electrostatic coalescer, dehydration tank and heater treater.

Electrostatic coalescers are very effective in achieving high dehydration efficiencies. They use electrical fields to promote coalescence of small droplets of water-in-crude into larger droplets.

The large droplets settle in the water layer at the bottom and drained continuously.

For dehydration tanks, wet crude enters through an inlet spreader and is distributed across the horizontal cross-sectional area. The oil gradually rises up in the tank and flows into a collecting device. The water droplets fall against the upward flow of oil, forming larger droplets and settling into the water layer at the bottom of the tank. Dry crude and water are continuously pumped out separately.

A heater treater uses fire tube to heat wet crude. The heat promotes the dehydration process. The rise in temperature reduces the viscosity of the oil that assists the migration of water droplets, settling into the water layer and oil floating on top. Dry crude is pumped out from an upper outlet and water drained from a lower.

Desalting

The salt content (known as salinity) of some crude oil may be very high and hence need to be desalted to meet customers specifications. Desalting is carried out by mixing heated crude oil with water in a valve that ensures that much of the salt content is dissolved by the water. The mixture of salt-in-water and crude oil is sent into a separator vessel. In the separator vessel, the salt-in-water solution settles on the floor and is continuously drained off. Desalinated crude oil floats on top and pumped out for further processing or shipping to customers.

Crude Oil Stabilization

Crude oil types that are sour are further processed to remove hydrogen sulfide contaminants in stabilizer columns. Booster pumps push oil to the top of the stabilizer column and it flows down via series of trays. On its way down the oil is channeled via reboilers that heat it up. This helps to boil off gases contained in the crude and with it hydrogen sulfide and light hydrocarbon. The gases rise to the top of the stabilizer and the heavy crude oil settles at the bottom. The rising gases also do the work of preliminarily heating the sour crude and subsequently sent to flare. The hot stabilized crude (now referred to as sweet crude) is passed through heat exchangers to reduce the temperature and then pumped to storage tanks for shipping to customers and refineries.

De-Oiling Facilities

Water separated during dehydration contains some quantity of crude oil. Some of this oil needs to be removed and the final quality of the water depends on the disposal method and the local legislation. There are two common methods of disposal: discharge to surface waters and injection into subsurface reservoirs.

There are various techniques for removing oil from production water, viz: gravity separation, gas flotation and hydrocyclone.

There are three common types of gravity de-oiling facilities. They are skim tanks, API oil interceptors and plate interceptors. Skim tanks work on the same principle with dehydration tanks. API interceptor is a pit into which the oily water flows, the floating oily layer is removed by a skimming device and de-oiled water leaves the pit over a weir. The plate interceptors use plates arranged in parallels and installed inside API-interceptor type of pit to improve the efficiency of the separator.

Gas flotation technique is sometimes deployed in series with gravity separation if the quality of the water from the latter is not sufficient. In this process the oil droplets are removed by attachment to rising gas bubbles. These oil bubbles rise and float on the surface of the oily water like scum and skimmed off.

Hydrocyclones make use of centrifugal forces to effect separation of the oil droplets entrapped in produced water. Oily water is pumped into a cylindrical swirl chamber at high velocity, setting off vortex with a reverse flowing central core. With the fluid accelerated through the concentric reducing section of the equipment, the separation takes place due to the difference in density between oil and water. The lighter oil moves towards the lower pressure central core and removed via an orifice at the center of the inlet header. The cleaner water flows out through the downstream section.

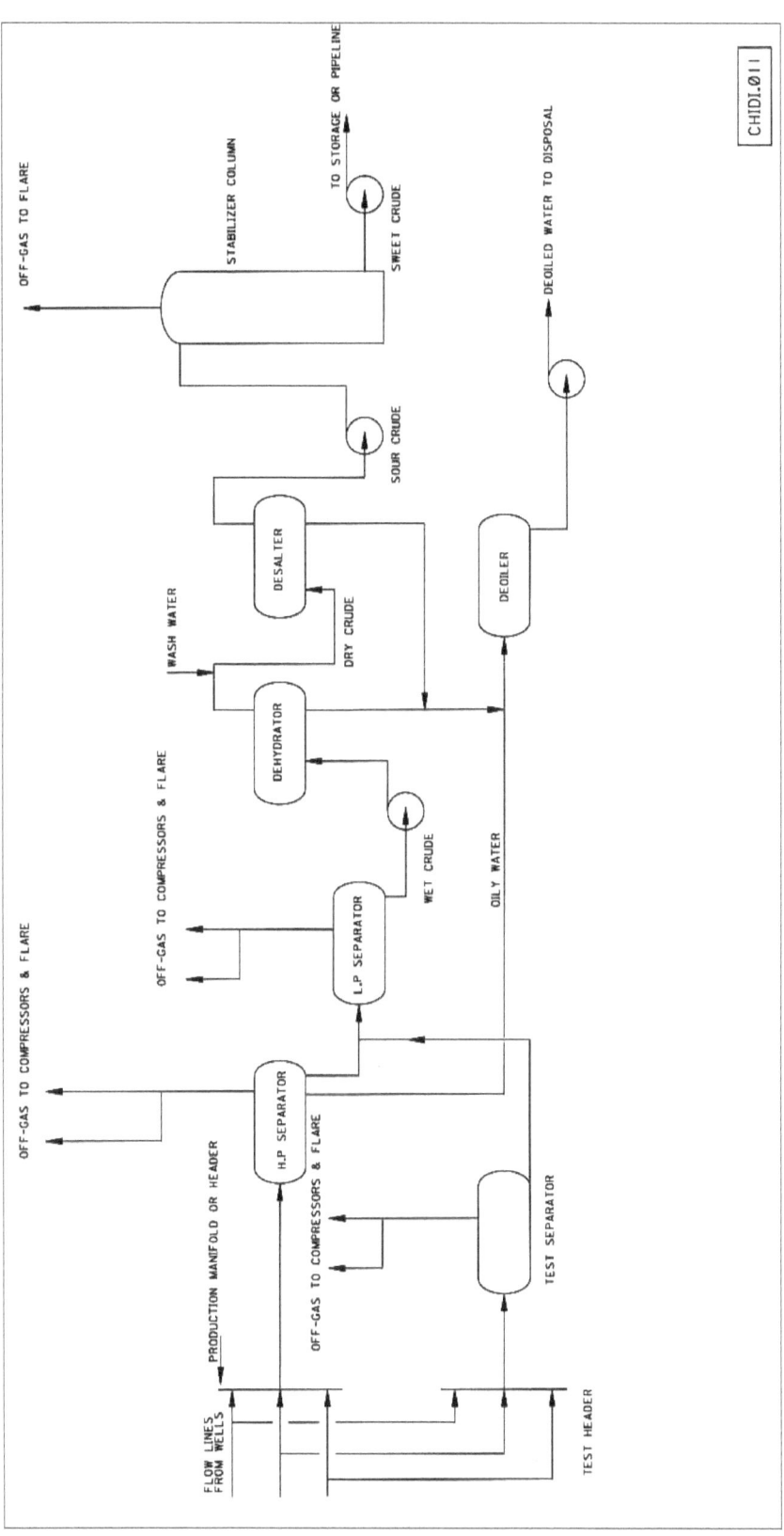

Figure 1: Simplified Schematic of Gas/Oil Separation, Dehydration, Desalting & Stabilization Process

CHIDI.Ø11

Gas Compression

The pressure of gas that is removed at the separation stage is normally too low for further processing and hence need to be increased (by compression). Before entry into the compressor suction, the gas is passed through a knock-out drum. Depending on the design, the drum uses some sort of baffles or demister pads to remove liquid carryover in gases and hence protect the compressor. Large liquid carryover have been known to damage compressors.

Compression of gas takes place in at least two stages, low and high pressure compression with intermittent cooling. The cooling process lessens the power required for compression, and ensures that heavy hydrocarbons entrained in the gas are dropped. This in turn increases the oil yield and keeps the compressor discharge temperature within manageable limits.

For gases that are rich in ethane, butane, propane and other heavier hydrocarbons, known as natural gas liquids (NGL), it is important to recover them from the gas. This is done by passing the process gas through stripper columns and either pumping the heavier hydrocarbons back into the crude oil line or sold to customers. Ethane is a feedstock for the petrochemical industry, while propane and butane (known as liquefied petroleum gas, LPG) are used for domestic purposes.

Gas Dehydration

It is important to remove water in the liquid phase from the gas. If left, it might cause hydrates and corrosion where carbon dioxide and hydrogen sulfide exist. There are many ways of gas dehydration but the most common method is by tri-ethylene glycol (TEG) contacting. Wet gas enters the bottom of the TEG contactor column, and lean TEG (about 98% pure) is passed from the top. Inside the column, wet gas passes through mist pads, rises up and gets dehydrated as it comes into contact with the lean TEG coming down. Dry gas then leaves from the top into a knock-out drum to remove any carryover TEG. Rich glycol is collected in a tray at the bottom of the column and piped off to be regenerated.

Rich TEG from the bottom of the contactor and that removed from dry gas at the knock-out drum are sent for regeneration into lean TEG and re-used.

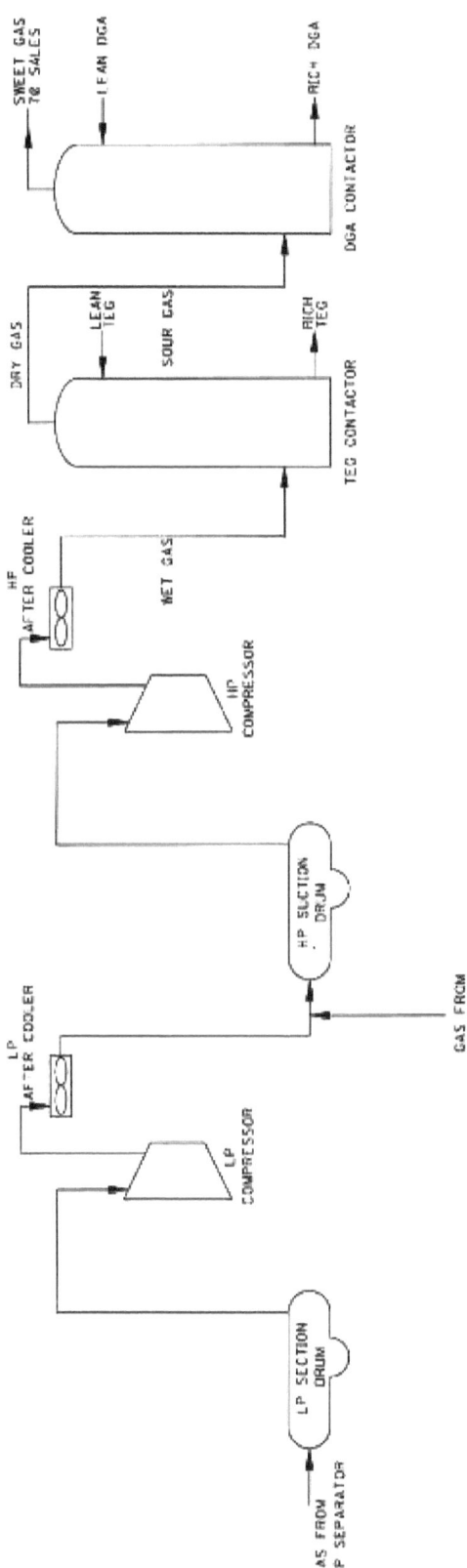

Figure 2: Simplified Schematic of Gas Compression & Treatment Process

Hydrogen Sulfide Removal

Hydrogen sulfide is required to be removed from gas before further processing. Otherwise a slight exposure of humans to very small quantities of sour gas can lead to fatality. Treating the gas to remove hydrogen sulfide is done in DGA (Di Glycol Amine) contactor. Sour gas is fed through the bottom of the Contactor while lean DGA is fed to the top and as the gas rises to the top, the lean DGA absorbs the hydrogen sulfide in it. The sweet gas leaves from the top of the column and sent out for sales.

The rich DGA gets out from the bottom of the contactor and is sent to the DGA stripper. Acid gas (with about 35% hydrogen sulfide content) is stripped out from the rich DGA and sent out from the top of the column to sulfur recovery modules. The rich DGA is regenerated to lean that can be used again.

Sulfur Recovery

In the sulfur recovery modules, acid gas is sent into a furnace and it is burnt with hot air. The products of this combustion are hydrogen sulfide and sulfide dioxide. Catalyst is used to get the hydrogen sulfide and sulfur dioxide react to form elemental sulfur that is then removed via condensation. This is followed by multiple other catalytic/condensation stages in-order to increase the percentage of sulfur recovery. After much of the sulfur is recovered, the tail gas is burnt in a thermal oxidizer and discharged into the atmosphere as sulfur dioxide.

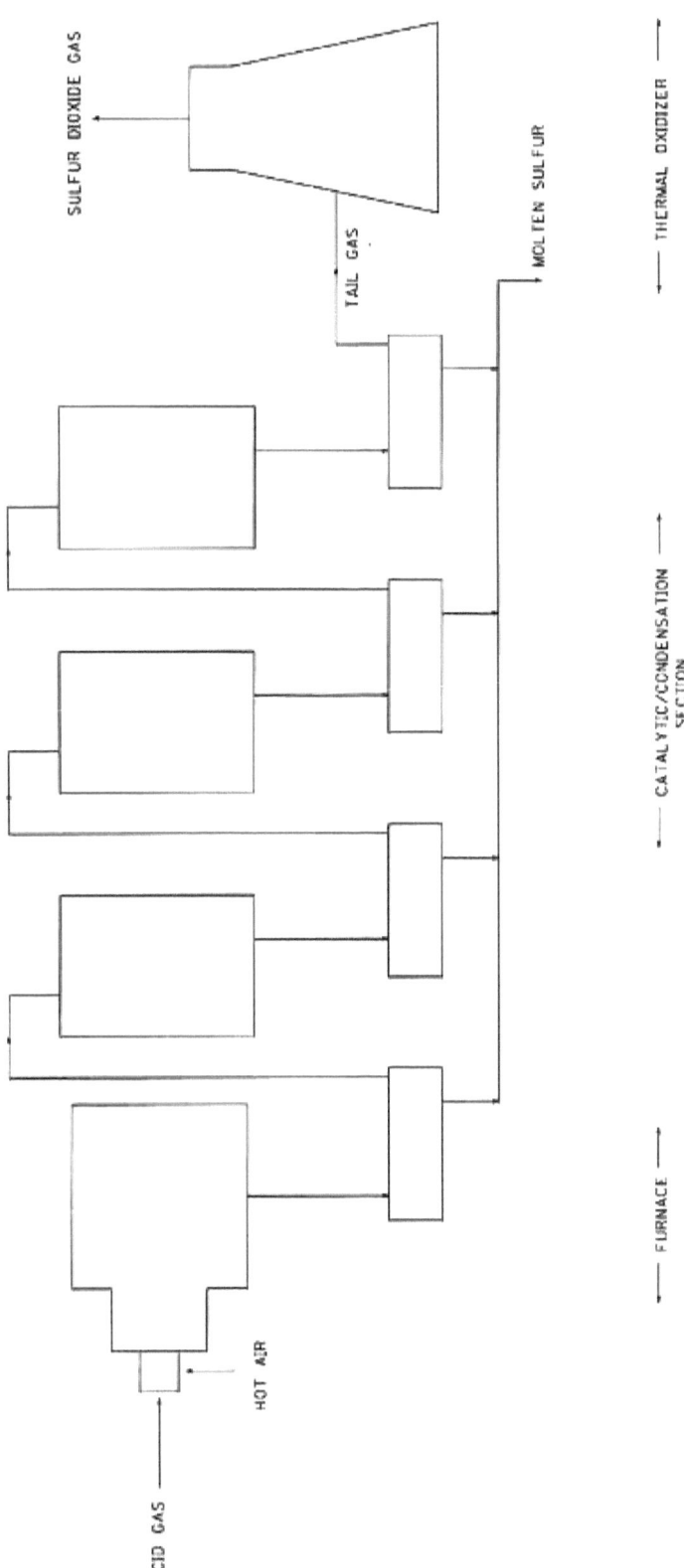

Figure 3: Simplified Schematic of Sulfur Recovery Process

Crude Oil Refining

Configuration of crude oil refineries vary. However, the most common refining process basically involves distilling in a column. The column has set of trays arranged at different levels. The characteristics of the various constituents of crude oil to boil at different temperatures is exploited in separating them. The liquid oil is heated to a vapor and rises upward in the column. The top is cooler than the bottom and hence the rising vapors (of the various constituent chemicals) condense at different levels at the respective trays. Light products (like butane) rise to the top, followed by gasoline, naphtha, kerosene, diesel and heavy gas oil in that order. The residue settles at the bottom.

Some refiners convert heavy liquids into lighter ones in a process known as cracking. For example, gasoline is in more demand in several parts of the world so there is an incentive to crack heavier hydrocarbons to make gasoline. This involves breaking long hydrocarbon molecules into smaller ones in cracking units separate from distillation columns.

In some cases, there could be a need to raise the quality and volume of gasoline produced. This makes use of catalysts, with series of reforming processes, to convert substances into aromatics and isomers that have higher octane numbers than the ordinary paraffin and naphtha. This rearranges the naphtha hydrocarbons to create gasoline molecules. Such reformate with higher octane number burn cleaner and hence more environmentally friendly. Another product of the catalytic reforming process is hydrogen that is used in hydrotreating.

Hydrotreating process is mainly done in refineries that process sour crude oil in-order to remove contaminants such as sulfur and other heavy metals. This is done by binding the contaminants with hydrogen, and passing through separate columns to remove them. The resultant byproducts are sold to other industries.

The final part of the refinery is the blending section, where various products are mixed to meet customers specifications. For example certain percentages of reformate, cracked gasoline, octane enhancers, anti-oxidants, anti-knock agents, rust inhibitors, metal deactivators and detergents may be blended to produce the gasoline you are familiar with.

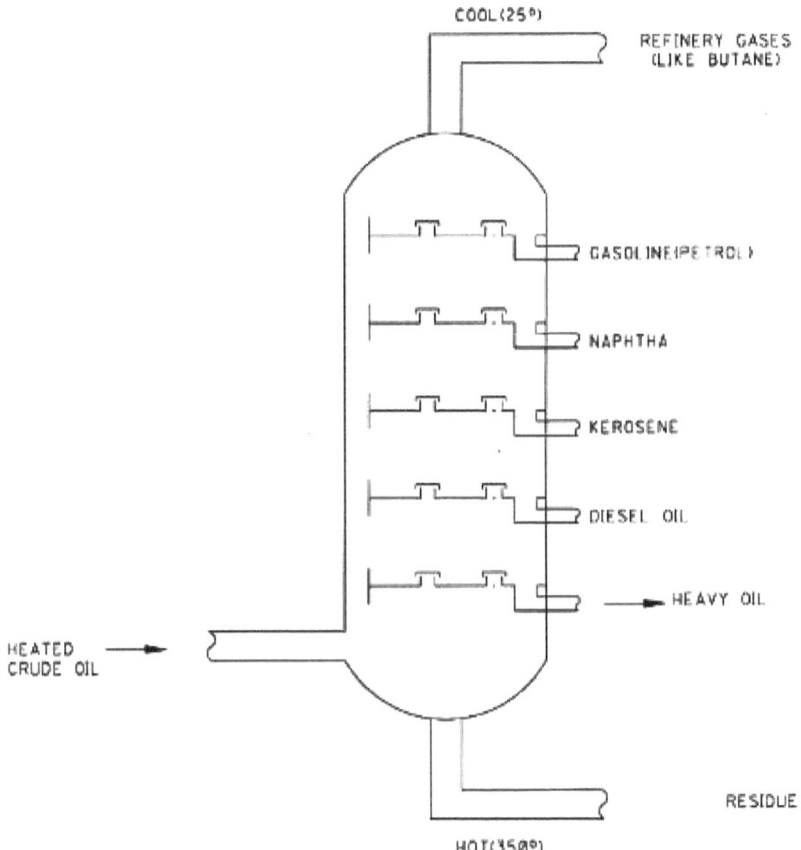

Figure 4: Crude Oil Distillation Column

Utilities Facilities

Utilities like electricity, water, steam, and compressed air (as a minimum) are required to run petroleum process facilities. Depending on country and geographical location, these systems may be connected from the public supply or generated within the facility. In some installations, sweet hydrocarbon gas serves as fuel in turbines that generate electricity.

CHAPTER 3

PROCESS HAZARD ANALYSIS TECHNIQUES

Process hazard analysis (PHA) basically involves gathering necessary hazards information, assessing them and making appropriate decisions. The analysis helps to discover what could potentially cause a major accident (hazard identification), the likelihood that it would occur and the potential consequences (risk assessment) and the available options for preventing and mitigating the consequences of such an occurrence.

Putting controls in place to prevent process accidents are expensive. However, it is much more expensive to do nothing and the accident happens. Carrying out proper PHA gives confidence that:

- The hazards in the process are understood;
- The most appropriate and cost-effective hazard control measures are identified;
- The control measures will effectively address the hazards and effectiveness will always be maintained.

In order to conduct a good PHA, the following information must be compiled: the hazards of the highly hazardous chemicals used or produced by the process (and non-process hazardous materials used), the technology of the process, the equipment and environment where the facility is sited. These are called process safety information.

Information on the hazards of the chemicals should consist of at least the following: physical properties, reactivity, corrosivity, toxicity, permissible exposure limits, flammability, thermal and chemical stability.

Information on the technology of the process should consist of the following: process flow diagram; process chemistry; safe upper and lower limits of temperatures, pressures, flows or composition (and consequences of deviations off the safe ranges); maximum intended inventory; and plot plan.

Information on the equipment should consist of the following: materials of construction, piping and instrument diagram (P &IDs), electrical area classification, relief system design and design basis, ventilation system design, standards and codes used in the design, material and energy balances, safety systems (such as detection, suppression and interlocks).

Environmental information should consist of the following: climatic condition, geotechnical data, neighboring facilities, nearness to population/communities, and topography.

The following are some of the methodologies for carrying out PHA, using the above information.

- Hazard Identification Study (HAZID);
- Hazard and Operability Study (HAZOP);
- Checklist Analysis;
- What If Analysis;
- Failure Mode and Effects Analysis (FMEA);
- Bow-Tie Analysis/Safety Case Approach;
- Layers of Protection Analysis (LOPA).

Each methodology has its niche. For example, while HAZOP is effective in assessing process operations, FMEA addresses equipment systems. The particular methodology chosen depends on the intent of the analysis and expected outcome. Sometimes, a combination of methods is used, for example, Checklist/What If, HAZOP followed by LOPA, HAZID followed by Bow-Ties/Safety Case etc.

PHA is a team work and the success is dependent on the skill and experience of the facilitator. He or she needs to maintain focus and keep energy levels high in order to produce the desired result.

Hazard Identification (HAZID)

HAZID is mostly carried out at the conceptual design stage as soon as process flow diagrams, draft heat/mass balances and plot plans are ready. It is meant to identify the main potential hazards and evaluate the severity of the consequences of the release of such hazards. When identification and assessment of such hazards are carried out early it provides critical input to project decisions at a stage when design changes have minimal cost implication. In some instances, projects have been cancelled outright because of unacceptable risks discovered during the HAZID study.

In addition, HAZID study has as one of its output a hazard register that summarizes the hazards present in a facility with their locations, sources, significance and controls. The hazard register, which is the essential first step in effective process hazards management, is a regulatory requirement in some countries.

Before the commencement of the study, all information about the process materials, the non-process hazardous materials, and equipment (process, utilities and other ancillary) proposed to be used should be gotten. Also required are information on existing site infrastructure, neighboring facilities/communities, weather and geotechnical data in order to adequately identify potential external hazards.

During the HAZID session, the criteria for screening the hazards should be established. The plant or facility should then be divided into sections. In each section, the potential hazards inherent in some parameters are identified by applying the guide words or checklist against each parameter. Some of the parameters could be:

- Materials inventory (raw materials, intermediate products, byproducts, finished products or fuel);
- Equipment or process;
- External factors or environment;
- Equipment spacing or layout;
- Safety or protective equipment or systems.

Some of the guide word or hazards checklist could be:

- Fire,
- Explosion,
- Toxicity,
- Extreme temperature,
- Noise,
- Radiation,
- Vibration,
- Asphyxiation,
- Corrosion,
- Erosion,
- Reactivity,
- Electrocution,
- Mechanical Failure,
- Carcinogen.

For example, for each material type (either raw materials, intermediate product, byproducts, finished product or fuel), the guide word or hazards in the above checklist will be applied and checked to reveal any potential hazards. If for example, a check reveals that a certain material is highly explosive, the study team could recommend some barriers or mitigation measures. The recommendation could be a complete substitution. Where enough information is not available, it could be recommended that further analysis be conducted on the material.

A typical HAZID study team should be between 3 to 6 personnel and should involve the design engineers, project management, commissioning and operations

Hazard and Operability Studies (HAZOP)

HAZOP is a structured, detailed and rigorous methodology used in the chemical industry in identifying how a process may deviate from design intent. It looks at the design intent of the process and assesses what might happen if there is a deviation from that intent during operations. The essence of the HAZOP study is to identify potential problems that present safety risks and efficient operation and not necessarily resolving them. It is usually conducted at the design stage but is also valuable in existing facilities in order to give assurance that major hazards have been identified, assessed and addressed.

A HAZOP study is carried out by a team of experienced personnel from various disciplines. It is conducted by proceeding step by step through the process and assessing each relevant process parameter. The key process parameters that are assessed are; FLOW, TEMPERATURE, PRESSURE, LEVEL and COMPOSITION. At each step of the process, guide words are applied to a parameter to identify potential deviations from the design intent. The basic guide words are; NO, LESS, MORE, PART OF, AS WELL AS, REVERSE, OTHER THAN. When deviations are identified, potential consequences are assessed, likelihood estimated and protections are proposed (if they are obvious). If the protections are not obvious, recommendations can be made to look for a way to address. The study team should not be bugged down trying to resolve every identified concern. If the HAZOP is on an existing facility, existing protections should be evaluated for effectiveness.

HAZOP Team Composition- A typical HAZOP team composition for a medium to large process facility should consist of the following members (small facilities can be done by lesser team, as required):

- Team leader/facilitator – he should be trained, and experienced in leading HAZOP studies. He should not be responsible for making any major technical contribution and needs not to have in-depth knowledge of the process and design. In fact, it is preferable that he does not have any knowledge of the subject so that he does not take any preconceived position, thereby allowing some hazards to fall through the cracks;
- Scribe – should be experienced in taking records of HAZOP studies;.
- Operations/process engineering;
- Maintenance (mechanical, electrical and instrumentation);
- Operations;
- Inspection;
- Loss prevention/safety;
- Other specialties as needed.

HAZOP Step-By-Step – For a good HAZOP study to take place there should be updated reference materials, the key of which is the piping & instrumentation diagrams (P & IDs). For new projects, it is important to have the following as well; process flow diagrams (PFDs), plot plans, design basis scoping paper, equipment data sheets, operations/ maintenance manuals, and materials safety data sheets (MSDS). The key steps in carrying out a HAZOP are as follows:

- With the use of the P&ID, the facilitator should section the process to be assessed into discrete nodes for which a specific and clear design intent can be described;
- For a selected node, the operations/process engineer should describe the design intent, specifying relevant process parameters;
- The facilitator should then select a process parameter and apply a guide word to build a deviation. If for example, the node is a pipe spool piece that the design intent is to transport specific quantity of process fluid per unit time from one vessel to another, the guide words could be "more flow";
- The team should then brainstorm on the possibility of "more flow" than the design intent. They should determine if there could be undesirable safety consequences of having "more flow";
- If undesirable consequences do exist, they should identify the potential causes for the deviation (more flow) and risk rank the potential consequences;
- The team should identify existing protections, determine if additional protections are required and recommend as appropriate;
- They should select the next guide word (e.g. "no flow"), apply it to same process parameter and repeat the above steps. This should be repeated for all relevant guide words, for all relevant process parameters and all nodes until the whole facility under study is completed.

Checklist Analysis

Checklist analysis is basically used for systems and processes that are covered by standards, codes and industry practices. The checklists are created by using the applicable standards, codes or practices to generate questions with which deficiencies in the facility under study can be identified. They should be prepared by personnel experienced in the systems or processes. After they are prepared, the analysis is undertaken by the team touring the area, and identifying hazardous situations.

'What-If' Analysis

'What-If' Analysis is the least structured of PHA techniques. It is carried out by a group of personnel experienced in the process brainstorming on potential deviations or failures. During the analysis, questions like "What if the tank overfills?" are framed, a scribe records all questions (probably on flip charts or sticky notes) and they are then grouped. The study team then splits into groups to handle various groups of questions.

For each question, the team should identify the accident scenarios, consequences and existing protections. They assess the effectiveness of the existing protections and possibly suggest further protection or alternatives. The technique can be used in identifying deviations from design, construction, modification or operations.

For simple systems, one or two people can carry out the analysis while larger teams should be assigned to complex facilities. The study can be for the entire process in a facility or may focus on specific consequences of concern.

Sometimes 'What-If' is combined with the checklist methodology. The 'What- If' helps to identify hazards and accident scenarios while the checklist has the advantage of more detailed systematic approach.

Failure Mode and Effects Analysis (FMEA)

A Failure Mode and Effects Analysis (FMEA) focuses on equipment systems (like mechanical processing equipment, control systems, protective systems). It is used in assessing how an equipment system fails and the consequences of such failures.

The inputs to a FMEA are an equipment list and P & ID. The assessment team should comprise personnel with good knowledge of the equipment, failure modes and effects of such failures. During the sessions the team should assess; for each item of equipment, a list of failure modes, the consequences of each failure, effectiveness of existing safeguards and recommend additional protections (if required). The outcome of the implementation of FMEA recommendations should be increasing reliability and safer operations.

Safety Case/ Bow-Tie Analysis

The safety case/bow-tie analysis approach to the management of major accident hazards in the process industry existed since the late 60s in the nuclear industry. However, after the Piper-Alpha incident it was made a regulatory requirement in the UK. As at today it is required by regulation in the UK, France and Australia.

It provides a documented demonstration that risk reduction philosophies and measures have been developed and implemented to ensure that the risks are tolerable and as low as reasonably practicable (ALARP) through the systematic application of the hazards and effects management process (HEMP).

A typical safety case, aside the introduction, has the following four key parts:

1. Facility description – this should contain:
   o A detailed description of the facility, including plant layout, material selection, safety systems, process systems and utilities;
   o List of major accident hazards identified and all safety critical elements or hardware barriers (as identified during the bow-tie analysis).

2. Organization and safety critical tasks – this should contain:
   o Normal operation facility manning levels and listing of key positions;
   o An organization structure, showing all personnel that hold safety critical positions;
   o In tabular form, safety critical positions aligned with their safety critical tasks, where the tasks fit into the bow-ties,

a brief description of the tasks and link to implementation procedure and performance standards/specifications.

3. Hazards and effects management process (HEMP) – this is the core of the safety case and is carried out by a team of experienced personnel just as you would have in HAZOP. It should contain the following; risk register, bow-tie analysis and manual of permitted operation (MOPO).

   o Risk register - In this section, all hazards in the facility should be identified and their risk on people, environment, the assets and the organization reputation assessed, using a risk assessment matrix. At the end of the assessment, for low or medium risk hazards the controls are discussed and then added to the risk register. Hazards that are assessed as being high or very high are modeled further using the bow-tie methodology.

   o Bow-tie analysis – this is a model that presents how a hazard may be released, escalated, and how it is controlled. When properly carried out, it contains the elements required to effectively manage a hazard such that the risks are tolerable and ALARP. For each hazard with the potential for high or very high risk, the bow-tie methodology should: identify the hazard release, escalation and consequence scenarios; identify barriers and categorize the barriers into either design (inherent) features, safety critical element, operational safety processes or operational intervention tasks; recommend activities/tasks required to maintain the barriers; and ensuring that these activities/tasks are linked to management systems.

   The role of a barrier on a bow-tie diagram is to prevent the release of a hazard or limit the consequences if the hazard is released and major accident occurs. It should be an independent protection layer (IPL). Examples of barriers that are categorized as design (inherent) features are separation distances, minimization of leak sources, reduction of process pressures etc. Examples of safety critical elements are pressure relieve valves, process containment systems, gas detection systems, alarms etc. Operation safety processes are things like work permit system, lock-out/tag-out etc. Some examples of operational intervention tasks are plant monitoring, alarm response, shutdown etc.

For every high or very high hazard that is assessed, the team should demonstrate all applicable measures to reduce risk to tolerable and ALARP levels have been developed. This can be based on regulatory systems, global standards, company standards, best practices, equipment/product data, or engineering judgment.

A bow-tie has the advantage of clear visual representation that makes it easier to conduct the ALARP review. In addition, it helps in incident review process if such a major incident occurs.

There are several proprietary softwares for conducting bow-tie analysis such as 'THESIS' and 'Bow-Tie XP'. The softwares have advantage of prompts that ensure the loop is completed for every assessment. Figure 5 is an example of a bow-tie diagram.

o   Manual of permitted operations (MOPO)– this specifies the operating envelope and safe operating limits for a facility. It also gives guidance on required action in the event of deviation from these envelopes such as adverse weather conditions, simultaneous operations, unavailability of a safety critical element or unavailability of critical manpower.

4. Improvement action plan, which lists actions required to address all concerns raised during the development of the safety case and also contains brief description of how the case is continually improved through the use of annual safety action plans.

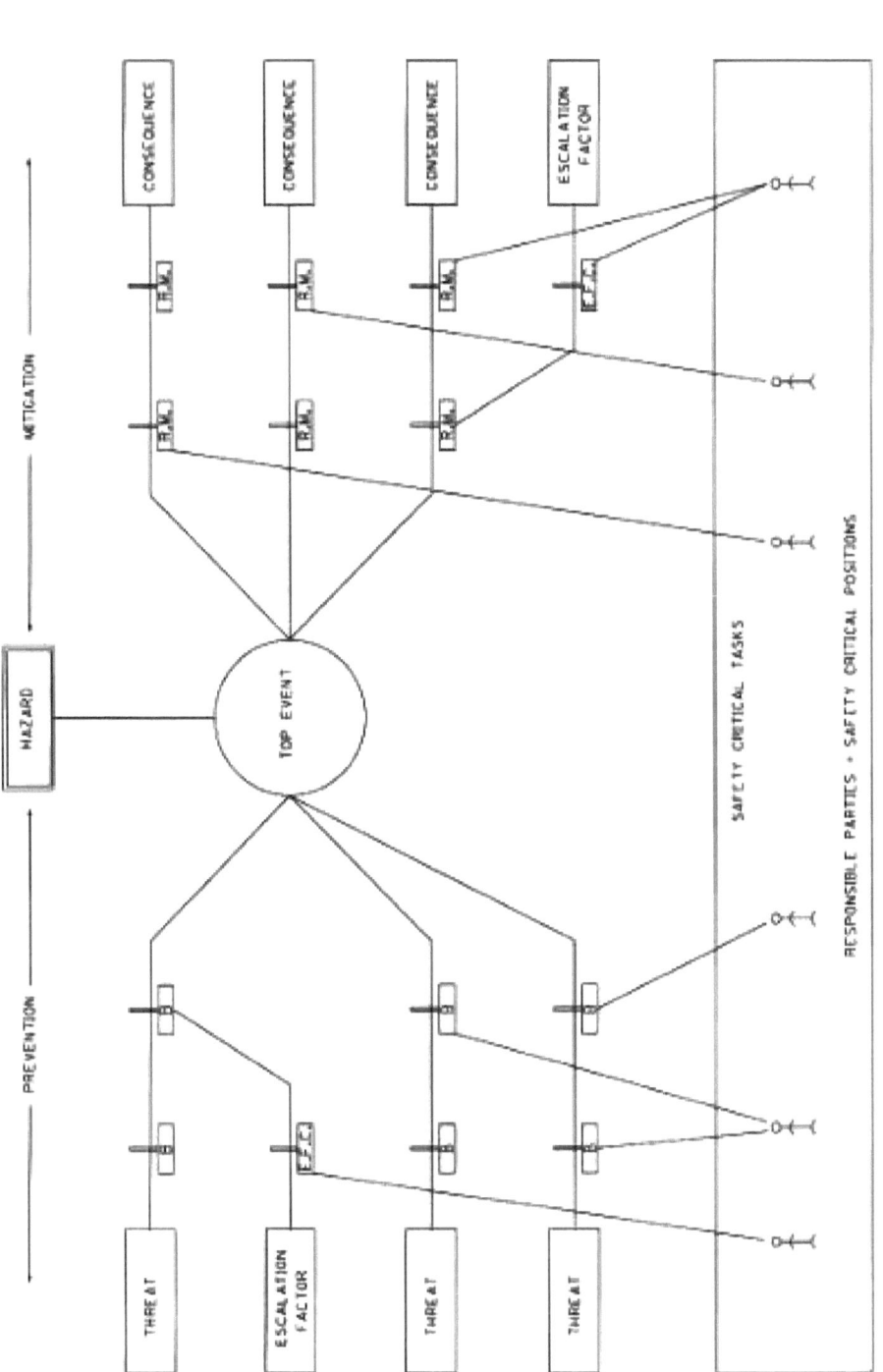

Figure 5: Simplified Bow-Tie Diagram (B being "Barrier"; EFC, Escalation Factor Control; RM, Recovery Measure)

Layers of Protection Analysis (LOPA)

In the process industry, the various measures for prevention of major accidents and mitigation of the consequences are known as 'layers of protection' (LOPs). The purpose of these layers is either to prevent the threats from releasing the hazards that develop into incidents or to mitigate the consequences of incidents if they occur.

The layers of protection are grouped into:

- Inherently safer design;
- Basic process control systems;
- Process alarms, fire and gas detection systems;
- Pressure and vacuum relief systems;
- Emergency shutdown systems;
- Drainage and containment systems;
- Control of ignition sources;
- Emergency evacuation systems;
- Fire suppressions systems.

Figure 7 is a bow-tie diagram that illustrates the relationship between initiating events or threats, layers of protection (LOP) or layers of defense (LOD), top events or incidents and consequences.

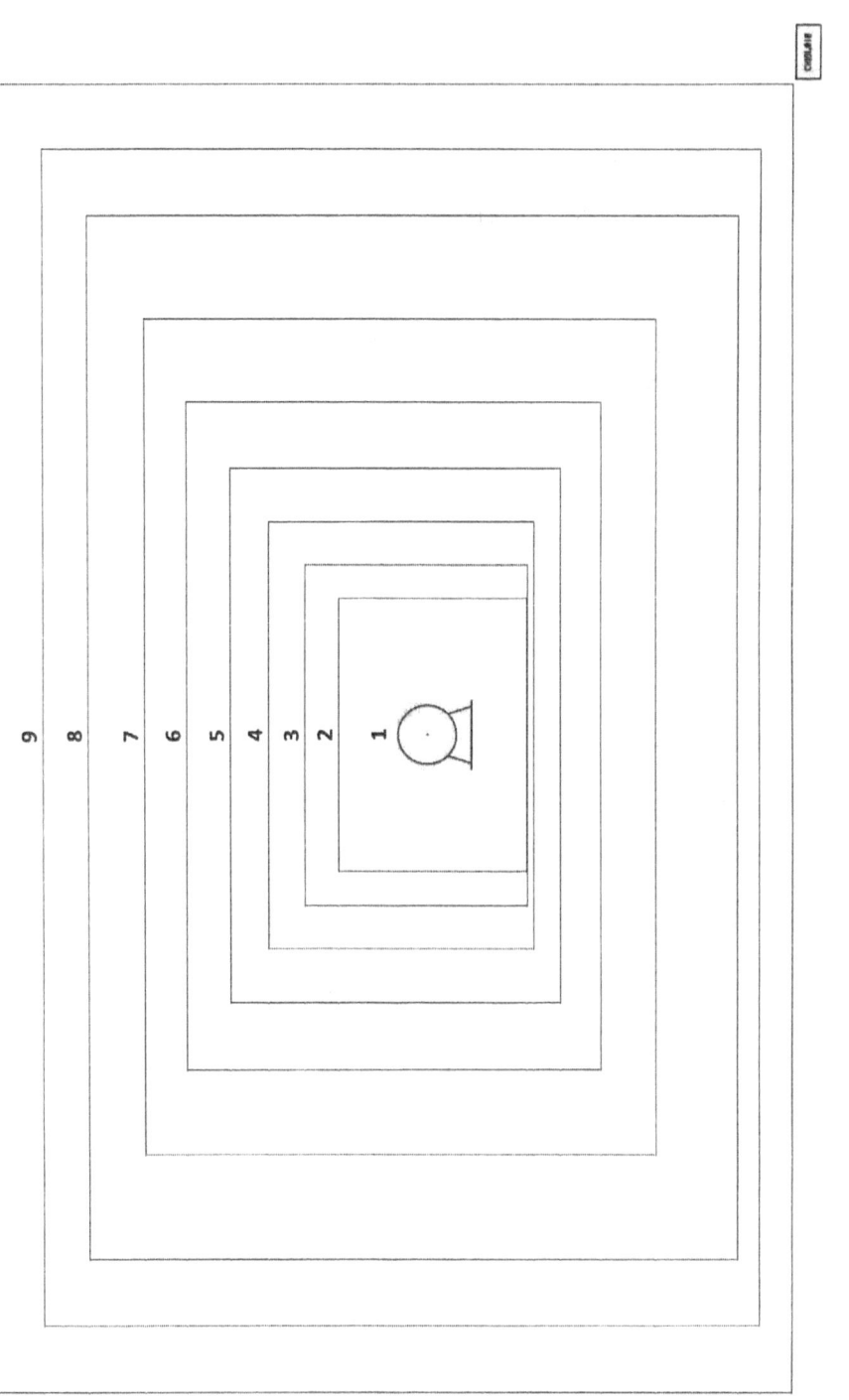

Figure 6: Layers of Protection (1, being "Inherently Safe Designs"; 2, Basic Process Control Systems; 3, Process Alarms, Fire & Gas Detection Systems; 4, Pressure & Vacuum Relief Systems; 5, Emergency Shutdown Systems; 6, Drainage & Containment Systems; 7, Control of Ignition Sources; 8, Emergency Evacuation Systems; 9, Fire Suppression Systems)

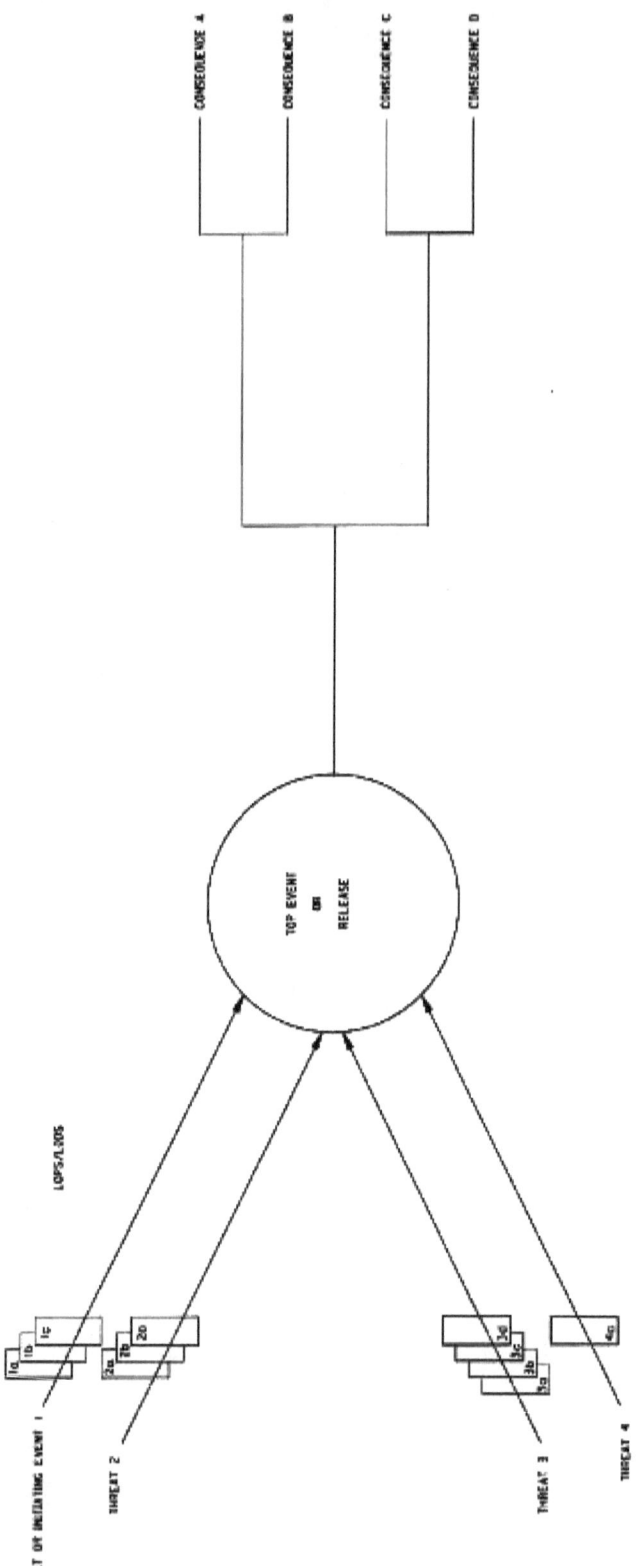

Figure 7: Bow-Tie Demontrating Relationship Between Initiating Event, Layers of Protection, Top Event (Release) and Consequences

As can be seen in the above bow-tie:

- An incident can be as a result of any of several threats or initiating event;
- There are usually more than one LOP designed to prevent a threat from developing into a top event;
- One LOP/LOD may serve to prevent several threats;
- A top event can develop into several consequences depending on the LOPs deployed and their effectiveness.

Layers of protection analysis (LOPA) is an analytical tool for assessing the adequacy of protection layers to ensure that process risk has been mitigated to as low as reasonably practicable (ALARP). It is a semi-qualitative risk assessment tool normally used after HAZOP to identify safeguards that meet the independent protection layer (IPL) criteria.

For protection layer to be independent, it must meet the following criteria:

- Specificity: is capable of detecting, preventing or mitigating the effects of a specific and potentially hazardous event(s);
- Independence: is independent of all other protection layers associated with the potentially hazardous event. This implies that the performance is not affected by the failure of any other protection layer or the condition that caused the failure of another protection;
- Dependability: it can be depended on to reduce the identified risk by a specified quantity;
- Auditable: it can be periodically verified that the protective function is still effective.

Some examples of IPL are pressure relief valves, safety instrument systems, gas detection systems, containment dikes, blast walls, and deluge systems.

The reference required for LOPA are: HAZOP report; equipment data; P&IDs for the facility; process description and philosophy documents; operation and maintenance manuals; cause and effects charts; and plant layout drawings.

The major steps of conducting LOPA are as follows:

1. The study team should list all process deviations and hazard scenarios with high or very high risk ranking. Each scenario has two elements; an initiating event (this is the element that starts the main chain of events) and a consequence (this is the result if the chain of events goes without any safeguard). If a single initiating event leads to multiple consequences, this should be taken note of and each consequence assessed in different scenarios;

2. For each process deviation, the frequency of each initiating event should be determined. The frequency should be gotten from credible sources like national, industry or manufacturer's databases;

3. The next step would be to identify the protection layers and estimate the probability of failure on demand (PFD) for each layer. There are industry databases where the PFDs for every type of device can be gotten;

4. The mitigated frequency would be the product of the initiating events frequency and the PFDs of the independent protection layers (IPL);

5. The mitigated risk should then be estimated by plotting the consequence versus the mitigated frequency;

6. This mitigated risk should then be evaluated for acceptability by comparing it with the organization's tolerable risk criteria;

7. If the risk remains unacceptable, other IPLs should then be recommended;

8. These steps should be repeated for all scenarios and all initiating events.

For better understanding of the layers of protection, each layer will be discussed in later chapters.

CHAPTER 4

INHERENTLY SAFER DESIGNS

Inherently safer design concept is the first layer of protection in a process facility. It permanently reduces or completely eliminates hazards or reduces the consequences of incidents arising from their release. This is a concept that involves considering several options that include reduction of a hazard, elimination of a hazard, substitution with a less hazardous material, using less hazardous process conditions, or designing a process to reduce the potential for human error, equipment failure or intentional harm or consequences of incidents arising from them. These options are considered from the design stage and throughout the facility life-cycle. It is worthy of note that no option will guarantee absolute safety. But a design is inherently safer if it reduces the risk and the reduction is permanent.

There are four broad strategies of achieving inherently safer processes, viz:

- Minimize – this means the reduction or minimization of the quantity of hazardous materials or energy in a process facility. This involves minimization of stock of raw materials, in-process intermediate storage tanks and use of innovative new technologies that lessen the length of time hazardous materials are in the facility;
- Substitute – this involves the substitution or replacement of a hazardous material or process with a less hazardous alternative. For example, a catalyst can be used to enhance chemical reaction

at a less hazardous process condition (e.g. lower temperature and pressure). Substitution strategy can be utilized not just in raw materials but also in materials of construction, insulation materials, heat transfer media and transportation containers;

- Moderate – this strategy has a thin dividing line with substitution. It means using hazardous materials under less hazardous condition (e.g. by dilution);
- Simplify – this strategy means designing a facility that is devoid of unnecessary complexity. Some complexities could be eliminated if PHAs are carried out early at the design stage when changes would easily be made. Otherwise, the design could end up being avoidably complex (e.g. with extra controls, alarms, safety instrumented systems, etc.).

Plant Siting, Layout And Spacing

The terms 'siting' and 'layout' can sometimes be mixed up. While 'siting' refers to the process of geographical location of a plant, site or facility, 'layout' is the relative position of various equipment or units within a site. 'Spacing' is the minimum safe distances between equipment or units.

Plants or facilities siting should be carried out in line with local regulations. The siting should ensure that the facility does not constitute threat to local population, environment, public utilities, other nearby facilities and vice versa. Adequate environmental, safety and health impact assessments should be carried out in line with local requirements and recommended mitigations to address assessed potential consequences effectively implemented before decision is taken on siting of a facility.

When a site is selected, the layout and spacing should be determined by considering the following:

- High hazard operations;
- Type of process;
- Process flow and interdependency of facilities;
- Number and positioning of personnel at risk;
- Criticality of operations and/or importance for continuing operations;

- Similarity of operations;
- Natural drainage pattern (site topography);
- Ignition sources;
- Prevailing wind direction and speed;
- Climate and other natural hazards;
- Fire and explosion exposures;
- Exposure to incompatible materials;
- Access for maintenance and emergency services;
- Other plants, adjoining facilities and public transportation;
- Future expansions;
- Flammable materials storage.

Plant layout should generally be based on the flow principle such that process materials will follow the process flow diagram. The entire site should be divided into process blocks of similar hazards (e.g. process units, utilities, loading, storage, offices, maintenance workshops, inlet area, loading or discharge, flare etc). There should be adequate separation between blocks and separation between different units in the same block. The separation distances could be roads, gravels or non- combustible materials to act as fire buffer zones. The essence of fire buffer zones is to prevent the impact of fire or hazardous materials releases from one unit affecting other units, allow controlled drainage of flammable materials, provide personnel escape routes, facilitate access of emergency response vehicles and ease maintenance/repair activities. A process block can be further sub-divided into smaller blocks based on risk.

Process units are the areas where the highest risks are likely to reside. It is important that appropriate risk assessment (using special methodologies to model fire, explosion and gas dispersion scenarios) be carried out to determine the layout and appropriate separation distance between equipment in the process units and between the process unit and other units.

In order to separate potential sources of ignition from fuel, the electrical area classifications of process plants should be used in determining layout and spacing. Electrical area classifications describe potential areas in which hazardous atmosphere may exist. For every classified area, fixed electrical or any equipment that may constitute an ignition source must meet the classification criteria to be located in or around it.

Pumps and compressors handling flammable products should not be grouped in one single area, located under pipe-racks, air-cooled heat exchangers and vessels.

Large vessels and equipment that need frequent maintenance should be located close to the block boundary. The same applies to equipment like heat exchangers that need space for removal and installation of internals.

Critical equipment, which is equipment necessary for safe operation and shutdown of a facility during emergency, should be identified and given adequate separation. Such equipment include process control systems, utility air supply systems, electrical power plants, substations, emergency shutdown systems, fire water systems, main process block valves.

Occupied buildings should be adequately spaced from areas of high risk such that people are adequately protected from fire, heat radiation, blast overpressure, shrapnel, and hazardous vapor cloud. Control rooms and other occupied buildings that must be in close proximity to the process facilities must be constructed of explosion-proof walls and have positive air-pressure to avoid ingress of hazardous atmosphere.

Electrical sub-stations and motor control centers should be located away from hazardous areas. This should increase the reliability of these assets if a loss event does occur. Electrical distribution cables should also be buried in order to limit their exposure to fire, explosions, windy weather, and mobile equipment.

Workshops, warehouses, laboratories and office buildings should be located away from process areas because they could be potential sources of ignition. In addition, they will not be exposed in case of gas release, fire or explosion.

Loading racks, piers and wharves should be spaced away from the other areas to limit the risk posed by the large number of trucks, sea vessels, and rail-road cars carrying hazardous materials.

Tank farms should be located downwind of the other areas, and possibly at a lower elevation. The tanks should be arranged in rows of not more than two deep and adjacent to roads accessible to firefighting trucks. The spacing between them should be such that thermal radiation intensity from a tank on fire should not be able to ignite an adjacent tank. They should be appropriately diked.

There are two broad methods of determining separation distance in a chemical or petroleum process facility. The first method is to use generic separation tables from international standards, insurance companies or similar process plants. The second, which is complex, is to carryout fire/explosion and dispersion modeling studies involving the actual hazards and environmental conditions. The former is usually more conservative while the latter tends to be more accurate.

It is worthy of note that extensive spacing makes a process plant inherently safer but increases the initial cost of building it. However, this is very little in comparison to the loss prevention benefits.

Materials Of Construction

A facility design can be made inherently safer by the materials of construction. Failure of equipment material leads to loss of containment with often catastrophic consequences. To make a design inherently safer, the design engineer should understand the internal process and the ambient environment, and select materials with the right properties to withstand the hazards posed internally and externally. At the fabrication or construction stage, proper fabrication procedures should be implemented. These should be followed up with regular inspection and maintenance.

Civil And Structural Support Design

The safety of a facility may depend on the civil and structural support design. If foundations or a structure supporting an equipment or facility fails, this may lead to rupture, loss of containment and release of hazardous materials. A design can be made inherently safer if the civil and support structures are properly designed to withstand all foreseen failure probabilities.

CHAPTER 5

BASIC PROCESS CONTROL SYSTEMS

Basic process control system (BPCS) can be considered as the next layer of protection after inherently safer designs. It is a system that monitors and handles process parameters (such as flow, pressure, temperature, level) in order to achieve desired objectives. Otherwise there is potential for loss of containment of hazardous materials with catastrophic consequences.

Monitoring and control can be achieved by operations personnel checking the value of the process parameter (e.g. level of liquid in a vessel) and using his judgment to close the supply valve when the vessel has become full or open the valve when almost empty. This however is subject to all sorts of human errors. In today's petroleum industry, processes are controlled automatically, with no human intervention required. The basic process control system makes use of sensors, controllers, actuators and other necessary control elements to control system parameter(s) to achieve desired results.

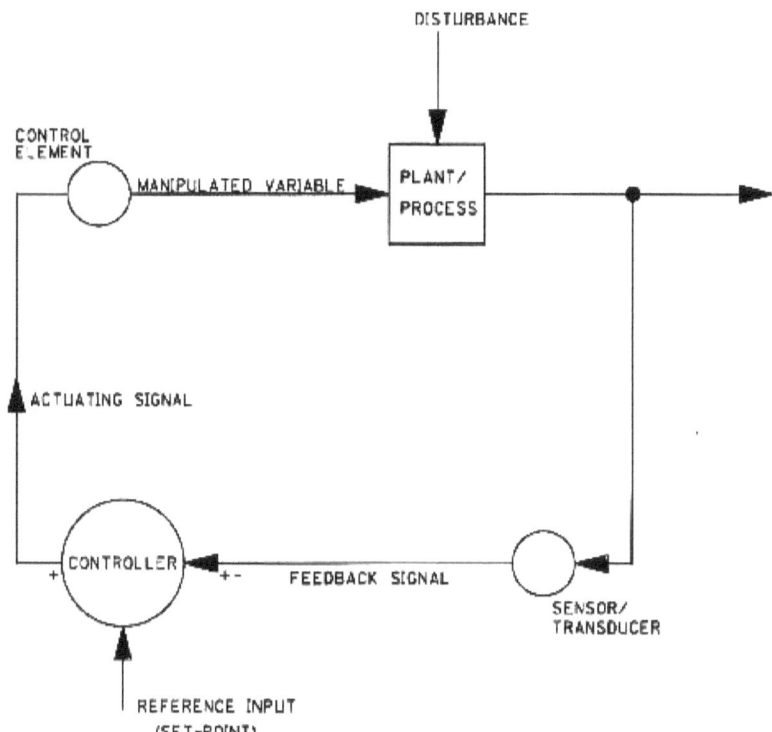

Figure 8: Process Control System

The above diagram gives a simplified explanation of how a process control system works. Suppose we have a plant or process and we are interested in controlling a variable (e.g. level or pressure). The control will be with reference to a set-point. The process variable is measured with the aid of a sensor/transducer (e.g. level or pressure transmitter). A sensor performs the measurement and converts it into analogous electrical or pneumatic information. The process of conversion is carried out by a transducer. Most times the transducers are integral part of the sensors. The information is subsequently sent into a controller that has an incorporated error detector device. The process variable is then compared with the set-point in the controller and an error signal with a specific quantity and polarity is generated. This error is further processed in the controller (e.g. actuator) and a control or actuating signal is sent to the control element. The control element (e.g. level or pressure control valve) initiates a physical change in the process variable by changing the manipulated value. This enables the process value to be kept at the desired value or set-point.

In a hydrocarbon process facility, there are tens or hundreds of process control systems monitoring and controlling process variables at different locations. These systems can be controlled from one central platform with the use of programmable logic controllers (PLC) and distributed control systems (DCS). The DCS does real-time video display of process variables of the various control systems and enables the operator to periodically intervene. The DCS segregates the controls for different areas of a facility, functionally and physically. Sometimes these segregated controls are housed in different buildings known as process interface buildings (PIBs). The essence of this segregation is to ensure that a malfunction or an incident in one part of a plant does not cripple the entire facilities.

Sensors/transducers need to be regularly calibrated in line with the manufacturer's instruction. Controllers and control elements need to be function-tested prior to putting them into service. Periodic testing and preventive maintenance program should be in place and effectively implemented to ensure that the performance standards are met continually. Testing may be conducted separately on sensors, controllers and control elements. The sensors may be tested at intervals of a few months to a year quite easily by inhibiting the controllers. Testing of the controllers and control elements may lead to real process manipulation and may be undertaken less frequently

CHAPTER 6

PROCESS ALARMS, FIRE AND GAS DETECTION SYSTEMS

Process Alarms

In petroleum process industries, an alarm system is a layer of protection used to notify plant operators and other personnel (via visual and audible means) of abnormal process situations or equipment malfunctions. It aids in the safe operation of the process under both normal and abnormal states. The primary function being to identify an abnormal situation, give warning signals early enough to allow remediation, guide personnel to the problem area, pinpoint the cause of the abnormality and give the required corrective action. When a corrective action is taken, the system should confirm whether or not it is effective. For an alarm system to be effective in discharging this function, it should in addition to being relevant to the operator's role, be clearly understood and presented at a frequency that the operator can handle.

The secondary function is to serve as an alarm and event log. This function helps the operator to analyze the events that led to current or previous abnormal process conditions. This function can be very useful in incident investigations and plant operation optimization.

Years back before the advent of DCS and other PC-based human machine interface (HMIs), alarms were selected with care because the panel boards have space constraints. In addition, the installation of the alarms required extensive work and when installed were expensive to change. But with the modern systems, installation, change and removal of alarms requires just a reconfiguration of the software. With this comes the temptation to have an alarm for every little deviation in the process or equipment malfunction. This has the potential to generate simultaneous multiple alarms that might overwhelm the operator. This situation is known as alarm flooding. Alarm flooding could lead to the inability of an operator to attend to a serious alarm with potential for catastrophic consequences.

To prevent this from happening, every process facility or plant that has alarms in the control system should have an alarm management system. Simply defined, alarm management consists of the set of documentations, procedures, practices, devices and methodologies that ensure that the alarm system in a facility is effectively managed throughout the life cycle of the facility. ANSI/ISA-18.2-2009, Management of Alarm Systems for the Process Industries standard specifies the requirements to create an effective alarm management system. There are ten stages in the lifecycle model of alarm management in line with ANSI/ISA-18.2-2009 and these are:

- Philosophy: this document defines the objectives and standards of the alarm system in a facility and the work processes and procedures to meet the specified objectives;
- Identification: At this stage, potential alarms are identified. If it is an existing facility, this can be gotten from the alarm database. Otherwise, this identification can be made with aid of critical operating limits, key process parameters, PHA/LOPA reports, P&IDs, PFDs, operating procedure and MSDS;
- Rationalization: This is the stage of ensuring an alarm is in line with the objectives and standards specified in the philosophy. This includes prioritization, classification, and determination of settings. The master alarm database and alarm design requirements are produced at the end of this stage;
- Detailed Design: The alarm system that meets the requirements produced above is designed. This includes design of HMI and advanced alarming techniques;

- Implementation: Here the alarm design is made live. This includes installation, commissioning, testing and training of operational personnel;
- Operation: At this stage the alarm is functional and refresher trainings are conducted, if necessary;
- Maintenance: This involves inspection, periodic testing, repair and replacement activities to ensure that the alarm remains reliable;
- Monitoring & Assessment: At this stage the alarm performance is monitored (on a continuous basis) and reported against the objectives/standards set in the philosophy. The reports generated could lead to changes;
- Management of Change: This process is to authorize changes (additions, modifications or deletion of alarms) based on monitoring and assessment reports;
- Audit: This is the periodic audit of the alarm management processes and possible recommendations for improvement.

It is worthy of note that these stages do not need to be followed in a particular order. If for example, nuisance alarms need to be rectified, the process could start from monitoring (monitoring and assessment stage) to identifying the culprit alarms. When the alarm is identified, we could go back to the records to cross-check the basis on which this alarm was selected (identification stage). The next stage could be to check that the alarm settings were properly defined and fixed (detailed design stage) and then check the basis for classification/prioritization (rationalization). After that we could play with the decision on how best to display the alarm (operation). When a decision is taken, authorization shall be gotten before changes are made (management of change). Then the change is implemented (implementation) and monitored again (monitoring and assessment) to ensure that action taken has eliminated the nuisance.

Fire Detection Systems

A fire detection system has the following objectives:

- Detect fire at the very early stage of its formation;

- Activate the alarm at the control room (giving the location) and at the location (to enable on location personnel to take necessary action);
- Activate the firefighting equipment (if so configured).

Generally, there are four main types of fire detection devices:

- Smoke detector;
- Flame detector, detects the ultra-violet (UV) or infra-red (IR) radiation emitted by the fire;
- Heat or thermal detector.

Each of these detection devices is unique in purpose and area of application. Hence the selection of type to use will be based on speed of response required, area of fire event and consequences to facility. An objective decision should be based on the recommendation of a loss prevention study. We will now look at each of these types of fire detection devices.

Smoke Detectors

There are various types of smoke detectors; the most common ones being the spot types (ionization and photoelectric), very early smoke detection and alarm (VESDA) system and reflected beam type.

The ionization (spot) type basically has a radioactive material, ionization chamber, an electrical source, positive and negative metal plates. The radioactive material generates 'alpha' particles that ionize the air in the ionization chamber. The positive and negative charges get attracted to the oppositely charged plates, thereby generating some electrical current between the two plates. When some smoke particles enter the chamber and interrupt the ionization, this will cause a reduction of the current and the alarm will be triggered.

The photoelectric (spot) type consists of a chamber, inside which is a light source, lens to gather the light into a beam, and photoelectric sensor. When smoke particles enter the chamber they scatter some of the light rays, directing them to a photoelectric sensor that activates an alarm.

Figure 9: Spot Type Smoke Detector

The VESDA system is a high sensitive type of smoke detector that works by drawing air samples via tubes from the protected equipment to a centralized detector which analyzes them for presence of smoke. This is used in very critical equipment such as electronic cabinets and cable ducting.

The reflected beam smoke detectors work on the basis of light obscuration. When there is smoke and it blocks part of the light beam the alarm is triggered. The latest technologies consist of a transceiver (a transmitter/receiver) and a reflector. The transceiver projects the light beam, monitors and receives it (when it is reflected by the reflector). The older technologies have the transmitter and the receiver as separate units (and no reflector). A beam type of smoke detector covers an area which would otherwise require a dozen spot types, thereby lowering cost of installation and maintenance. They generally have a maximum range of about 330 feet and a maximum spacing between detectors of 60 feet (in effect, a theoretical coverage area of 19,800 square feet) while the spot type has a maximum coverage of about 900 square feet.

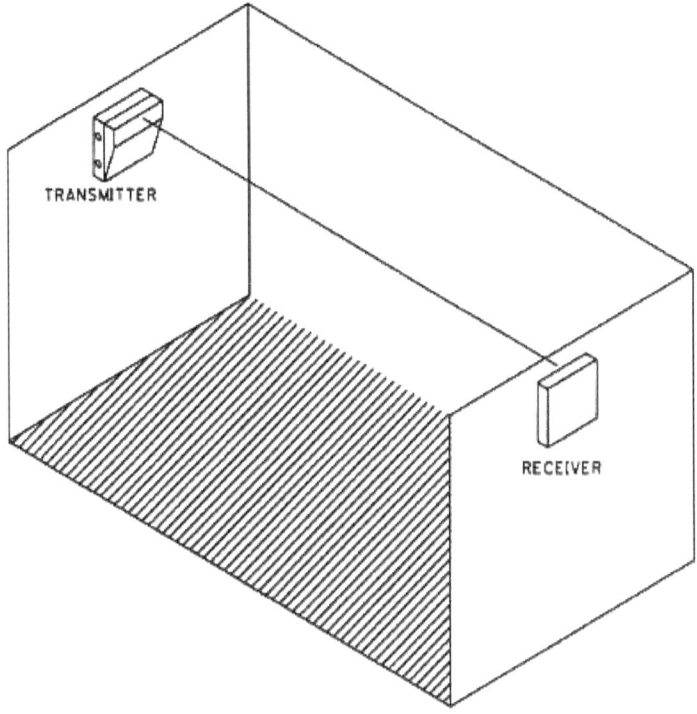

Figure 10: Schematic of Reflected Beam Smoke Detection System

It is worthy of note that smoke detectors are very sensitive and hence when used in dusty environments or smoky environments (like mechanical workshops and kitchens) will always be triggering false fire alarms.

Flame Detectors

Flame detectors are installed in areas where flame (rather than smoke) is the first part of fire that forms. There are three main types of flame detectors: ultra-violet (UV), infra-red (IR) and combination (UV and IR) types.

Ultra-violet (UV) type flame detectors work by sensing the UV light emitted from a flame. However, they could also be sensitive to a sun UV light, welding arcs, x-rays and lightning. Hence the use of UV flame detector should be used in facilities or equipment (like turbine enclosures) where there is no disturbance from sun, lightning, x-rays and welding arcs.

Infrared (IR) flame detectors work by sensing infrared radiation emitted by a CO_2 product of flame. It is more reliable than UV flame detectors because it is not sensitive to sun UV light, welding arcs, etc. and hence can be used in open areas or closed areas.

UV/IR combination flame detectors use both UV and IR sensors to detect a flame. Such detectors are configured to 2oo2 voting logic to activate an alarm. This minimizes incidents of false alarms caused by UV only or IR only.

Heat or Thermal Detectors

A heat or thermal detector makes use of a heat sensitive element (usually a thermistor) to detect fire. It has slow response because it activates only after the fire has gotten to some significant stage to radiate heat energy enough to be sensed by the heat sensitive heat detectors. It is used in areas like workshops, laboratories, kitchens, and smoking shelters where use of smoke or flame detectors has high potential for false alarms. There are three main types of heat detectors: fixed, rate of rise and fusible plug types.

A fixed type heat detector is set at a fixed temperature point and activates an alarm when the ambient temperature gets to the set point. It is suitable for facilities liable to swing temperatures like kitchens.

Rate of rise type of heat detectors work by sensing the rate of rise in the ambient temperature. This type of detector is very sensitive but cannot be used in facilities, like kitchens, with swing temperatures. It is suitable for use in places that don't have high air flow like warehouses and store rooms.

A fusible plug type uses a metal cylinder that has a sealed metal (with low melting point) and connected to a pneumatic tubing loop. When the ambient temperature reaches the melting point of the seal metal, the seal metal breaks, leading to a pneumatic air leak. This leak is detected by a pressure switch that activates an alarm. In remote locations, the alarm could be configured to also initiate firefighting or emergency shutdown systems.

Inspection, Testing & Maintenance of Fire Detection Systems

Fire detection systems provide early warning signs of a fire emergency, provided they are properly made, installed and regularly inspected and maintained. It is important that they are UL listed or FM approved and properly installed by certified and reputable alarm system companies. After installation, the system should undergo an initial acceptance testing. In addition, each time there is a modification and re-installation, there should be a re-acceptance test.

At least on semi-annual basis, visual inspection of a fire detector should be conducted to verify that the conditions of the detector at the time of acceptance have not changed such as to affect the performance. Such inspection should check for accumulation of dirt, physical damage, blockage, painting of the parts of the device etc.

Performance testing should be conducted at least annually by qualified personnel in line with the equipment manufacturer's specification. Adequate records of initial acceptance or re-acceptance tests, regular visual inspections and performance tests should be kept.

The National Fire Protection Association (NFPA), Standard 72, National Fire Alarm Code, specifies the standard for inspection, testing and maintenance of fire detection systems. However, local or national fire codes may have stricter standards.

Gas Detection Systems

A gas detection system has the following objectives:

- Detect the presence of combustible, flammable or toxic gases and oxygen depletion;
- Activate the alarm at the control room (giving the location), and at the location (to warn personnel at location of the hazard and giving them opportunity to escape);
- Activate the fire and gas system to control and minimize the leak (if so configured).

Gas detectors are either portable or fixed. The portable detectors are worn or carried around by personnel and are used to monitor the atmosphere around them. Fixed detectors are installed around areas where there is potential for gases to be present.

Gas detectors make use of sensors, some of which include: infrared point, electromechanical, semiconductor, and ultrasonic sensors. We will briefly describe the principles behind each of these types of sensors and their common applications.

Infrared (IR) point sensors: Infrared point sensors make use of ability of some gases to absorb IR radiation at certain wavelengths. In principle, it consists of two IR light sources and an IR light detector that measures the intensity at two different wavelengths. One is at the absorption wavelength and the other, outside the absorption wavelength. If a gas is present, it intervenes between the source and the detector in the absorption range of the sensor and hence the level of radiation reaching the detector is reduced. The intensity in this wavelength is compared to that outside the absorption range and the difference is proportional to the gas concentration. IR point sensors have remote sensing capability and do not have to be placed into the gas to function effectively.

Electromechanical sensors: Electromechanical sensors work by measuring the concentration of a gas by oxidizing the gas at an electrode and measuring the generated current.

Semiconductor sensors: Semiconductor sensors work by the chemical reaction that takes place when the monitored gas comes in contact with the sensor. The electrical resistance in the sensor decreases and this is used to calculate the concentration of the gas.

Ultrasonic (acoustic) sensors: These types of sensors function by detecting the changes in the background noise of the environment. If gas under pressure leaks, the generated noise is at much higher frequencies than the background noise and the sensor is able to distinguish these thereby triggering an alarm.

Testing & Calibration of Gas Detectors

Gas detectors should be function tested and calibrated regularly according to the manufacturers' instruction. Typically, a portable detector should be function tested on the daily basis due to the fact that it's being moved from place to place (and into different environments) could affect the effectiveness and functionality. Calibration of both portable and fixed detectors should be on a quarterly basis

CHAPTER 7

PRESSURE AND VACUUM RELIEF SYSTEMS

Pressure and vacuum relief systems are designed to prevent catastrophic failures of vessel and equipment due to overpressure or vacuum build-up. Over pressure or vacuum can be caused by any of the following scenarios:

External fire – If a vessel is exposed to heat input from external fire, causing thermal expansion or decomposition or vaporization this will lead to rise in pressure;

Operational errors – if normally closed valves are left open during operation, it could lead to the release of high pressure fluid resulting in a vacuum being created. On the other hand, a normally open valve could be inadvertently closed in error resulting into build-up of pressure;

Blocked outlet – Fluid outlet pipes from a process equipment can be blocked (e.g. by leftover construction materials that are not properly disposed-off) after turn around maintenance activities. This may lead to overpressure conditions if effective pre-startup checks are not carried out. Blocked vent lines could result in the creation of vacuum in a vessel;

Equipment failure – Tube rupture or control system failure in equipment such as heat exchangers can result into overpressure. Sometimes, a control valve left open could cause fluid to flow from high pressure to lower pressure systems;

Process upsets – Loss of cooling, unexpected feed increase or decrease, problem with contaminants/catalysts, instrument and control failures could lead to run-away reactions that may result to excessive pressure increases;

Utility failure – The failures of utility systems such as cooling water, power, instrument air and inert gas are common causes of equipment over pressure.

Pressure Relief Devices

There are several devices that are used in overpressure protection but the most common are safety relief valves and/ or rupture discs. These devices are typically connected to discharge either to the atmosphere, to a containment vessel or to disposal systems (e.g. flare). The following are the different types of relieving devices and their characteristics.

Conventional spring loaded relief valves- In these types of devices, the bonnet, spring and guide are exposed to the released process fluid and hence are used in non-corrosive service. In addition, the setting of the valve is affected by the amount of superimposed back pressure expected and are used in systems where the built-up back pressure does not exceed 10% of the set pressure.

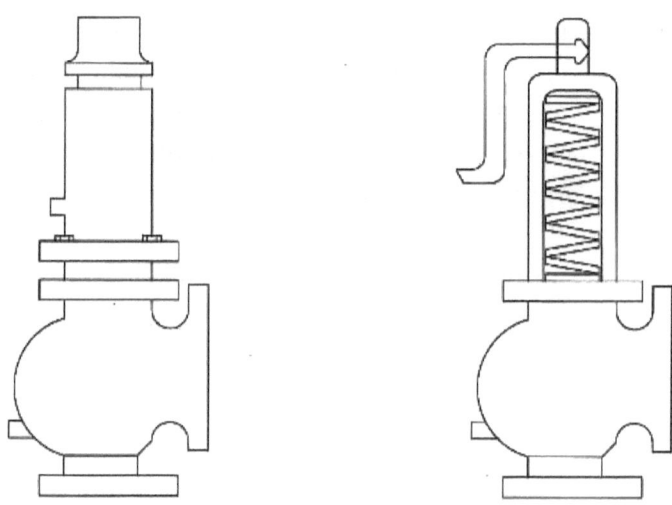

Figure 11: Spring Loaded Relief Valves

Balanced spring loaded relief valves – These can be used in corrosive systems as the bonnet, spring and guide are protected from the released process fluid. In addition, balanced spring loaded valves are not affected by back-pressure unless it rises to about 30% of set pressure.

Pilot operated relief valves – The pilot operated relief valve has the major relieve device (the main valve) that is controlled by a self-actuating pressure relieve valve (the pilot control unit). The pilot unit senses the process pressure and if it is up to the set pressure, it opens the main valve by reducing the pressure on top of an unbalance piston, diaphragm or bellows of the main valve. This will relieve the pressure until it gets to the blowdown pressure and the pilot closes down the main valve. It does this by increasing the pressure on top of the main valve.

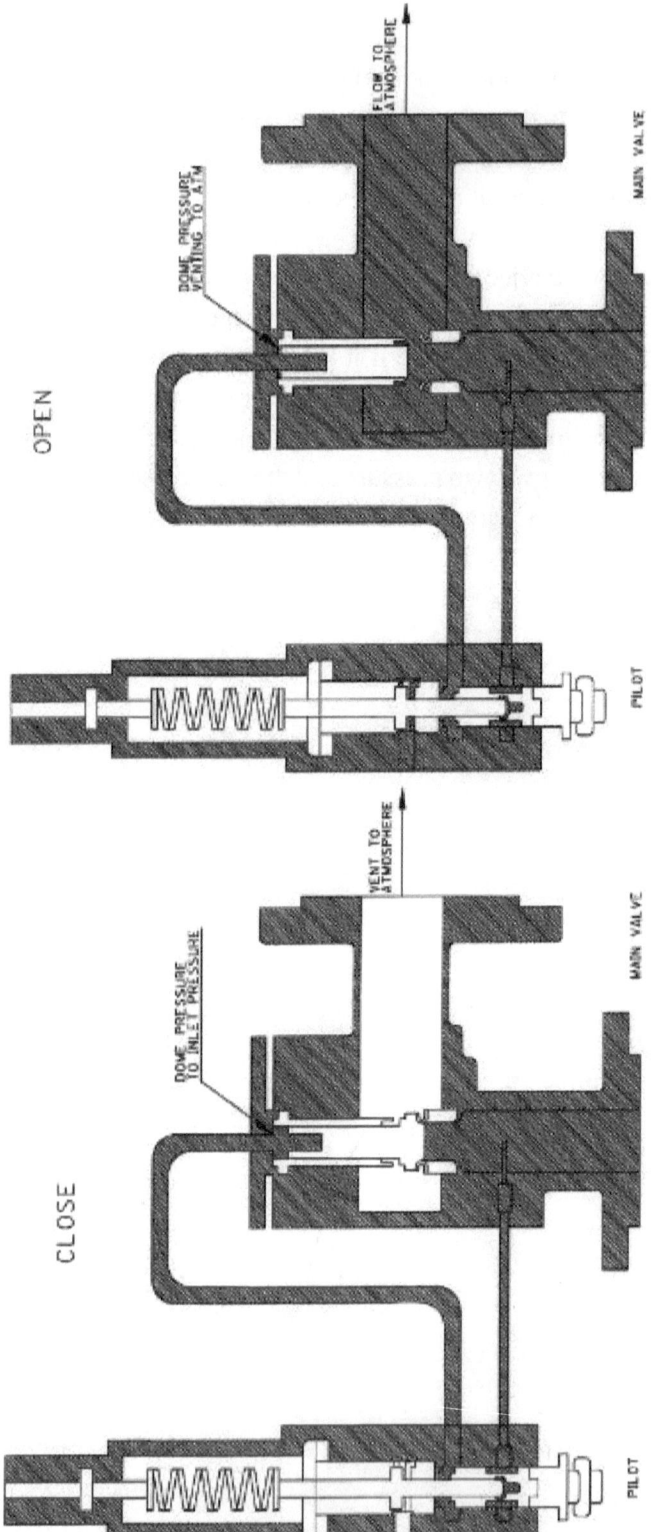

Figure 12: Pilot Operated Relief Valve

Rupture disks – A rupture disk is a non-reclosing pressure relief device that functions by the breaking of the pressure retaining disk when the inlet static pressure gets to a set level. The material of the disk can either be metal, graphite, plastic or a combination of materials. It can either be installed alone or before a pressure relief device to protect the latter from plugging. When it is installed before a pressure relief device, it is advisable to also have a pressure indicator to give an alert that the rupture disk has failed

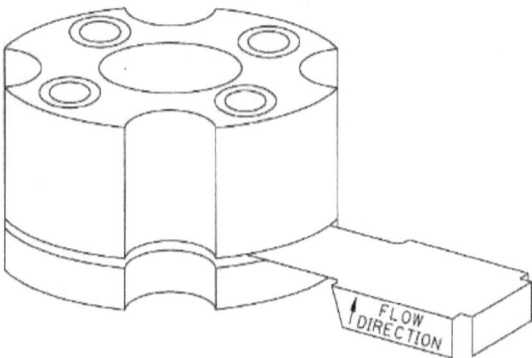

Figure 13: Rupture Disk

Liquid seals – These are u-tube hydraulic loops designed to be used in systems operating at slightly above atmospheric pressure. In most instances, they are meant to prevent ingress of air into an equipment operating close to ambient pressure, overpressure and vacuum. The loops are typically filled with water, oils or any other fluid that is compatible with the process fluid.

Pressure-vacuum relief valves – these are typically used in atmospheric and low pressure storage tanks. They combine both a pressure and a vacuum relief characteristics into one assembly.

Sizing, Inlet/Outlet Piping and Location of Relief Devices

If an overpressure occurs in an equipment with a relief device that is either not properly sized, piped or installed, there is the potential for failure to occur. It is critical that all relieving scenarios are adequately evaluated and the valve sizing and selection are correctly done.

The design for the inlet and outlet piping of the pressure relief devices is required to be done such that they can meet the mechanical and thermal stresses that develops when they relieve. It is not recommended to have threaded connections in high pressure system or vibrating equipment as they have the tendency of "unscrewing" while in service.

The location should be at the most advantageous point in terms density, temperature or phase for ease of relief and safe disposal.

Inspection, Test and Maintenance of Relief Systems

Relief systems are required to be inspected and tested before being taken into service by competent and authorized personnel or organizations. In addition, they must be periodically inspected, tested and maintained in order to ensure they operate properly. The frequency of test and inspection should be based on:

- Statutory requirements;
- Manufacturer's instructions;
- Test/inspection data from similar devices in similar processes;

- Operating history of the system - systems/equipment with frequent upsets that have actuated a relief device need to be inspected more frequently;
- Visual inspection results- for example, corrosion, signs of valve leakage or damage should result into bringing inspection, test and maintenance schedule forward;
- Position installed- for example, valves installed in a system with a common discharge header have the potential of being affected by discharges from other devices and that should increase the frequency;
- Shutdown window- it is ideally recommended to conduct inspection, test and maintenance using planned shutdown windows for other maintenance activities;
- Criticality of the system- systems that are critical to plant operation should be inspected more frequently;
- Unknown characteristics – if the characteristics of a device or the effects of the process system on it are unknown, it is recommended to make the first inspection not more than one year from installation. The result from this and subsequent inspections will gradually determine the best frequency.

When a device or system is tested, an immediate corrective action must be taken if there is an indication that it may not function properly. Record of each inspection, test and maintenance must be kept by personnel responsible for the protected equipment. In addition, a plate with inscription of test/inspection date and next due date should be attached to the device. The device must be inspected and tested before the next due date.

CHAPTER 8

EMERGENCY SHUTDOWN SYSTEMS

An emergency shutdown (ESD) system is a type of process control system that is designed to quickly stop the operation of a plant/ equipment, isolate it from incoming or outgoing flow of hydrocarbon, safely depressurize it and therefore lower the likelihood of an incident from happening or escalating. It is meant to protect people, equipment and the environment by minimizing the consequences of emergencies related to loss of containment. It should be in every petroleum plant/ equipment and without it, process incidents may be provided with endless supply of fuel that can destroy the whole facility. The system monitors a plant/equipment, responds to hazardous situations that can potentially destroy the facility and activates shutdown once certain parameters deviate from normal conditions.

Typically, the actions from an ESD system are as follows:

- Shut down the equipment or plant;
- Isolate hydrocarbon inventories & stop the flow;
- Isolate electrical systems;
- Depressurize and blow down the equipment or plant,
- Close fire doors (where applicable);
- Control emergency ventilation (where applicable);
- Prevent escalation of events.

Levels of Shutdown

There are several levels of shutdown. For every facility, the levels are specified at the design stage. Typically, these levels activate emergency actions for increasing areas of a plant/facility as the degree of hazard or involved areas increases. For example, for smaller incidents or smaller involved areas an equipment only may be required to be shut down and isolated. For larger incidents or larger involved areas an entire plant should be required to be shut down and probably de-pressurized quickly to prevent escalation. In situations where the shutdown of one equipment may present a hazard to another part of the facility both should be shut down. Typical levels of shutdown are as follows:

- Level 1 – Total plant shutdown (for potentially catastrophic incidents);
- Level 2 – Unit shutdown (severe incidents);
- Level 3 – Equipment shutdown (slight to major incidents).

ESD systems are in loops. For each equipment, unit, train or entire plant that has been assessed to have ESD, there shall be a loop. At the design stage of the facility, the ESD requirements should be specified, giving the ESD loops, functions, trip parameters, fail-safe position of final devices and the safety integrity levels (SIL).

The SIL specified for a system is the probability that it will act on demand and varies from 1-4 (the higher the number the higher the probability, implying that SIL-4 is more reliable that SIL-2).

Reliability and Fail Safe Logic of ESD Systems

An ESD system has to be independent of BPCS and the design of components based on fail-safe logic. The independence is derived by physically segregating process locations, instruments, impulse lines, logic devices and wiring of the ESD and BPCS. This ensures that failure of one does not affect the other. On the one hand, failure of the DCS should not stop the ESD from shutting down and isolating the facility. On the other hand, failure of the ESD system should not stop an operator from shutting down the plant using the DCS.

Fail-safe features are derived by selecting components in an ESD system such that in the event of failure of a component the process reverts to a safe condition. If a safe condition is to stop the flow of fluid, then the isolation valve shall be required to close. If the safe condition is to drain some hazardous fluid, then the valve shall be required to fail in open position.

Performance of an ESD system is measured in terms of reliability, availability and survivability. Reliability is the ability to perform under stated operating conditions for a specified time interval. Availability is what proportion of the time that the system will be available to perform when demanded. Survivability is how the system will perform post event (for example in a case of fire engulfing the facility, how long the system should continue to be effective). These are defined at the design stage, performance standards and criteria specified. At the construction phase, the specified criteria for the components of the system must be met. Routine maintenance, inspection and testing shall be implemented to assure that the performance standards are maintained.

Activation Mechanisms

An ESD consists of field-mounted sensors, valves and trip relays, system logic that processes input signals, alarms and activates outputs in accordance with the design. The input signals can be from fire and gas detection systems or from process instrumentation set points. There can also be manual activation points in the control panel and in strategic locations in the plant.

The manual activation points are required to be systematically arranged to ensure optimum availability and hence provide adequate protection to a plant or facility. They should be about 8 meters (25 ft) away from every high risk process hazard location and within 5 minutes walking time from any portion of the facility. The location should be at the prevailing upwind direction from the protected equipment, in the path of evacuation routes and near other emergency devices. It should not be in normal mobile equipment route or maintenance access so that they are not prone to frequent damage. The height of the activation switch should be such that it is convenient for all personnel.

The manual activation device should be a push button/knob and hardwired. In order to prevent accidental activations, a confirmed action of the button is required to manually activate the ESD system. This usually takes the form of a cover that needs to be lifted before exposing the button or knob.

It is required that the location of the activation point should be very visible, identifiable from a distance and clearly labeled, as to the equipment or area of coverage.

Final Output Mechanisms and Devices

For every ESD system or loop, there is at least a final output device or mechanism. There are two distinct classes of devices/mechanisms, emergency isolation valves (EIV) and energy shut-off devices.

Emergency isolation valves (usually ball valves) are used to stop the release of hydrocarbon (and other hazardous materials) when ESD is activated. An EIV has an actuator that uses power sources (such as pneumatic, electrical or hydraulic). When the ESD is activated, the actuator is energized and instantly acts on the valve to shut the flow of hydrocarbon into or out of the plant/equipment. Some actuators have fail-safe spring-return mechanisms such that in the event of the loss of power (from the above sources), the valve reverts to the safe position. Actuators that do not have spring-return mechanisms are required to have back-up power sources (like compressed air and hydraulic fluid accumulator drums or independent electric power supply source).

Energy shut-off devices shut the flow of energy (for examples, electricity, gas, steam etc) to an equipment or plant thereby safely stopping further release of hazardous material when ESD is activated.

In situations where the inventory in the plant/equipment could constitute more hazard if left in place, the ESD will include opening relevant EIVs to de-pressure and drain to a safe location (for example, to flare headers). This is applicable in catalytic systems with potential of exothermic run-away reactions and process side of process heaters.

Testing and Validation of Emergency Shutdown Systems

Emergency shutdown systems are required to be tested and validated regularly to ensure they will work when needed. They have to be trusted to function as expected during an unexpected event. The component parts may take years before being put to work. Within this period, they are subjected to degradation due to weather conditions, aggressive and corrosive environment. The question is whether or not they will still function when and as needed. This requires regular testing and maintenance in order to increase their reliability and integrity when called upon in an emergency.

The function testing of each ESD input/output shall include: operation/activation of field-mounted sensors; the programmed logic; alarms; manual activation buttons and activation of final output devices (like the actuators, valves, fire doors and emergency ventilation). These shall be done during factory acceptance test and pre-commissioning. They shall be repeated at frequencies prescribed by respective manufacturers or specified by user organization, whichever is stricter, and records of such tests kept.

As part of testing and validation, regular full stroke tests of EIVs are required. These are done during new installations, scheduled preventive maintenance and opportunistic break down maintenance activities. At these times, the valves are activated (fully open and closed) and checked for; tight shut sealing, air/hydraulic supply leakage on the actuator, stem blocking, valve/actuator set integrity, signal integrity on the control panel etc.

A partial stroke test can also be done. This means moving partially the valve shaft and taking measurements of the actions needed to do it. During this period the valve response time, possibility of blockage and pressurization of the actuator are also measured or checked. Partial stroke tests are more efficiently done using what is called valve intelligent positioner. This positioner measures the movement, does calculations and generates graphs showing the test result. Partial stroking may detect over 70 % of the problems while avoiding production loss due to shutdown (for full stroke test). There have however been instances where plants have accidentally shutdown while performing partial stroke tests.

CHAPTER 9

DRAINAGE AND CONTAINMENT SYSTEMS

In the petroleum industry, large quantities of flammable and combustible liquids are handled. In the event of loss of containment and ignition, these liquids form pool fires that can potentially damage exposed structures and equipment. Drainage and containment systems safely remove such spills rapidly and reduce the amount of fuel that could be involved. If there is no or inadequate drainage capabilities there will be no means of dissipating spilled hydrocarbon liquids other than being consumed by potential fire. Drainage systems play an important role in avoiding, reducing and preventing flammable and combustible materials spills from resulting into fire and explosion.

At the design stage of hydrocarbon facilities, adequate drainage capabilities should be provided around tanks, vessels, pumps, columns etc. The containment and drainage systems are required to be designed to safely contain and remove spilled liquid, firefighting water and storm water.

Drainage Mechanisms

For every equipment/facility that requires drainage system, one or a combination of the following mechanisms is used.

- **Grading**

In grading, the ground around the equipment is sloped appropriately to move the liquid in a desired direction in order to eliminate the possibility of the hazardous materials spreading to other areas of the facility or offsite locations. This mechanism can be used in both process and storage areas. The philosophy is to orient the slope away from high value or critical equipment to remote collection points such as oily water separation ponds and catch basins. This is sometimes supplemented by remote impoundment, diking, trenching or closed drains.

The ground in the graded area should be paved with concrete or asphalt so that the spilled liquid does not get absorbed into the soil, with environmental consequences.

Care should be taken to ensure that the drainage channel does not pass under other equipment and pipe racks. They should also not interfere with the use of the area for mobile equipment access and maintenance activities.

- **Diking**

Dikes are installed to contain, control and hold spills in storage areas (like tank farms) where there is potential for large spills in emergency situations. This type of containment is not used in process areas because the hydrocarbon would pool and constitute unacceptable risk to plant and equipment.

Dikes should be designed to be capable of holding over 110% of the volume of the tank around which it is built, if it is just a tank. Drain valves should be installed in-order to remove rain or fire water and prevent overflow of the dike.

The dike material should be non-permeable (for examples, concrete, solid masonry, asphalt, compacted earth) so that spilled liquid does not leak out to adjoining facilities. In addition, the walls should be designed to withstand the full hydrostatic head of the contained spill.

The height of the walls should be restricted to 6ft (1.8 m) for ease of firefighting, and to allow adequate ventilation in-order to prevent vapor cloud explosion. Provision should be made for properly sloped ramps through which firefighting vehicles can gain access in emergencies.

As much as possible, connecting pipes should be buried within the diked area to avoid failure when engulfed by fire.

- **Remote impoundment**

Impoundment basins (like oily water separation ponds) are located in a remote area of a facility where spills can burn safely, if ignited, with no risk to plant, equipment or buildings. The spilled material are directed to the remote basins via graded channels or trenches. The basins should be sized to contain the largest spill which cannot be readily isolated and fire water. Some basins have lift stations to skim and pump hydrocarbon back to a process equipment when the level at the remote impoundment gets to a set point to avoid overflow.

- **Trenching**

Rain, fire protection and other surface runoff water that have small likelihood of containing hydrocarbon are directed by grading to trenches. Trenches are located at the edges of process units and route non-oily water to remote impoundments. Certain percentage of hydrocarbon or other flammable liquids may still be mixed up with the water flowing through these trenches due to accidental discharges or blockage of closed drains. Hence they should not pass under or near process equipment, cable trays, vessels, pipe racks or fire water lines. They should be appropriately sloped to ensure continuous flow in the desired direction.

- **Closed drains**

Closed drains are the primary drainage mechanism in process areas. It mainly consists of catch basins, underground network of pipes, main headers, and central collection points. Spills, fire water and rain water are directed to catch basins (via grading) and carried through network of pipes/main headers to a central collection points. From the central collection points, the liquid is transferred to oil and water separation units. Catch basins are usually water sealed to prevent propagation of fire through the drainage system and combustible vapors from being released into the atmosphere (unless at the vent points).

Sealed closed drain systems should have vents to allow the relieve of trapped gases. If the gases are not vented, hydrostatic vapor lock will occur that will stop incoming liquids from draining into the system. The location of such vents should be appropriate in-order not to constitute hazards. It should be away from ignition sources, operating platforms or ventilating fan intakes, and high enough to be readily diluted by the atmosphere. The size of the vent pipes is required not be less than 4in (100mm) in diameter, especially in cold climates, in-order to prevent freezing and clogging. Appropriate warning signs should be installed around vent areas.

Closed drains should be designed to handle the largest anticipated flows, spill liquid, fire water, spent cooling water, "wash-down" water and rain water. It is important that network of pipes from incompatible liquids do not interconnect in-order to prevent undesirable reactions.

Inspection, Testing and Maintenance of Drainage Systems

For a properly designed and built drainage system, the major problems will arise from blockage by sand, debris and maintenance wastes. To minimize this happening, there should be good housekeeping, regular inspection and cleaning of drainage systems.

On a regular basis, a facility should be flooded with fire water and checked to see that it drains as quickly and effectively as designed. Areas of constraints should be identified, investigated and repaired. Repair could sometimes be simple de-silting, and removal of materials causing the blockage.

CHAPTER 10

CONTROL OF IGNITION SOURCES

A combination of an ignition source, oxygen and fuel (anything that can burn) leads to fire. Hydrocarbon is fuel, and there is abundant supply of oxygen in the air. If ignition sources are not identified and controlled in an atmosphere with the presence of hydrocarbon, the consequence is fire/explosion.

In the petroleum industry, any place where flammable/explosive atmosphere may occur in such quantity that can burn is referred to as a hazardous area. In such areas, ignition sources are effectively controlled by design measures, equipment selection and safe systems of work.

Sources of Ignition

The following are potential sources of ignition:

- Electrical/electronic equipment;
- Internal combustion engines;
- Hot surfaces;
- Static electricity;
- Lightning strikes;
- Hot work.

Hazardous Area Classification

Hazardous area classification is a method of analyzing and classifying potentially hazardous areas as per the type of hazardous materials (hazardous characteristics of the material and the risk). The purpose of this classification is to be able to identify the types of tools/ equipment that can be safely used in such areas. It also helps in assessing the type of activities that can be safely carried out. The understanding of this is key to control of hazards posed by ignition sources.

There are two main types of classifications, viz; International Electromechanical Commission (IEC) and North American. IEC presents its classification in Classes and Zones, while North American classification is in Classes/Divisions/Groups.

The IEC uses three main criteria in defining hazardous areas. These are the type of hazard (denoted in GROUPS), the likelihood of the hazard being present in flammable/explosive concentrations (ZONES) and the auto-ignition temperature of the material ("T" rating). Auto-ignition temperature of a hazardous material is the lowest temperature at which it will spontaneously ignite in a normal atmosphere without an external source of ignition, such as a flame or spark.

For GROUPS, we have:

- Group I - methane or gases/vapors of equivalent flammable range, mainly found in mines;
- Group IIA - propane or gases/vapors of equivalent flammable range;
- Group IIB - ethylene or gases/vapors in the same flammable range;
- Group IIC - acetylene or hydrogen or gases/vapors of equivalent flammable range;
- Group III - dusts, with several sub-divisions of that.

From the above, it is obvious that groups I and III are not common in the petroleum industry.

For ZONES, there are:

- Zone 0 – flammable/explosive atmosphere is continuously present or for long periods of time;
- Zone 1 – flammable/explosive atmosphere is likely present during normal operation;
- Zone 2 – flammable/explosive atmosphere is not likely to be present in normal operation and if it does, it will only be for short duration.

The above zones classification are for gases. Dusts are zoned as 20, 21 & 22 respectively.

Other areas outside these three zones can be classified as 'safe areas'.

For auto-ignition temperature (T rating), we have the following classifications:

- T1 – Ignition temperature of above 450 degree Celsius;
- T2 – Ignition temperature between 300 and 450 degree Celsius;
- T3 – Ignition temperature between 200 and 300 degree Celsius;
- T4 – Ignition temperature between 130 and 200 degree Celsius;
- T5 – Ignition temperature between 100 and 130 degree Celsius;
- T6 – Ignition temperature between 85 and 100 degree Celsius.

In the North American method of classification, flammable and combustible liquids are divided into CLASSES, as follows:

- Class I Liquids – Handled at temperature above the liquid's flash point and may produce a flammable atmosphere;
- Class II Liquids – Handled at temperature below liquid's flash point and do not produce sufficient vapors to form ignitable mixture;
- Class III Liquids – Normally do not produce vapors of sufficient quantity to form ignitable mixture.

The DIVISIONS are:

- Division 1 – Gases and vapors can normally exist (synonymous to Zones 0 &1 of IEC);

- Division 2 – Gases and vapors normally confined (synonymous to Zone 2 of IEC).

Use of Hazardous Area Classification to Control Ignition

With the knowledge of hazardous area classification of a plant/facility, the following are the ways to minimize or eliminate the risk posed by each of the ignition sources.

Electrical/Electronic Equipment

Electrical and electronic equipment used in hazardous areas are carefully selected to ensure that they do not arc or create sparks that generate heat that can ignite flammable/explosive materials. For an equipment to be located in these areas, one of the following control philosophies shall be incorporated in its design:

- Does not contain components that arc or create sparks;
- Has limited arcing or create negligible sparks such that the heat generated cannot ignite the specific hazardous atmosphere;
- If it cannot be prevented from arcing or creating big sparks, the heat generated should be prevented from being transferred to the hazardous atmosphere;
- The heat generated from arcing should be cooled before being released into the hazardous atmosphere.

To ensure that potential for fire or explosion is greatly reduced or even completely removed, electrical/electronic equipment that are used in hazardous locations are designed and manufactured (to meet one or two of the above control philosophies) under the below protection concepts (the designation 'Ex' meaning explosion protection):

Intrinsic Safety- Intrinsically safe equipment are designed and manufactured either as non-incendive or non-sparking. Non-incendive means that it can spark but does not generate enough energy to cause an ignition. Examples of intrinsic safe equipment can be found in two-way radios, cell phones, lap-top computers that are so designated. They have

'Ex i' marking on the equipment body. Type 'ia' (Ex ia) is meant to be used in zone 0 areas, type 'ib', zone 1 and type 'ic', zone 2.

Flame-proof – Flame-proof equipment are designed and constructed such that if there is a spark inside the equipment (e.g. junction boxes) and hazardous gas is also present (inside the junction box) and there is an explosion, the equipment (in this case, a junction box) is built to withstand the internal explosion. After the explosion, relief is provided via a flame path for escaping hot gases. The gap of the flame path is designed such that the hot gases are cooled along the path and when out they do not constitute ignition sources any longer. These types of equipment are marked 'Ex d'and are designed to be used on zone 1 locations of specific gas group and T-rating.

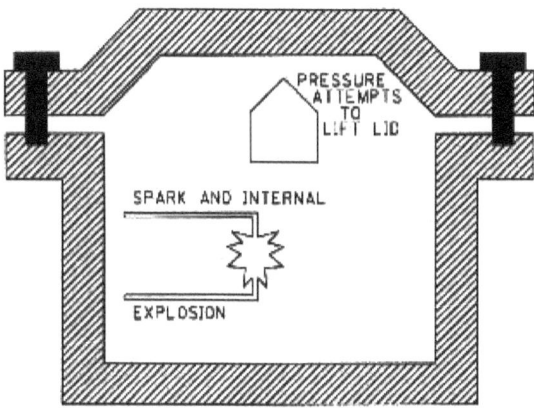

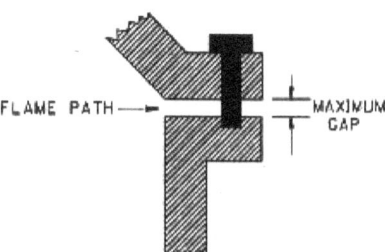

Figure 14: Flame Proof Equipment

Encapsulation – In this type, the equipment or the component of the equipment that has potential to cause ignitions is immersed (or encapsulated) within a compound or resin such that the spark or arc does not have contact with the explosive atmosphere. The compound or resin also has the function of limiting the surface temperature of the equipment under normal operating conditions. This type of equipment is identified by the marking 'Ex m' and can be located in zone 0 (if it is Ex 'ma') or zone 1 (if Ex 'mb').

Increased Safety – These are identified by the marking 'Ex e' and can be used in zones 1 &2. The normally sparking components are excluded and other components designed to substantially reduce the likelihood of fault conditions that could cause ignition by ensuring electrical connections are reliable, increased insulation effectiveness and reducing probability of dirt and moisture ingress. Examples can be found in electric motors, junction boxes and lightings.

Pressurized – These are marked 'Ex p' and can be located in zones 1 &2. Pressurized protection has two processes, one ensures that the pressure inside the enclosure (where there is the potential for sparks to occur) is sufficient to prevent the entrance of flammable gas, vapor or dust. The second process maintains a constant flow of air (or inert gas) to dilute any explosive atmosphere inside the enclosure.

Oil Immersion – The equipment components in this type of protection are completely submerged in oil. The oil provides an insulation to prevent ignition. Examples are switch gears and starters. They are identified by marking 'Ex o' and can be located in zones 1 &2.

Powder Filling – All equipment component that has potential to arc is contained within an enclosure filled with quartz or glass powder particles. The powder filling prevents the possibility of ignition. Examples can be found in capacitors, electronics, telephones and chokes. They are marked 'Ex q' and can be used in zones 1 &2.

Special protection – These are marked 'Ex s' and have no specific parameters. It is a method of protection that can provide a pre-determined level of safety to ensure no potential for ignition. It can fall under any one of the above methods or a combination of two or more.

In addition to having 'Ex' and the protection type inscribed on it, an equipment that has been designed for use in hazardous areas should also have indication for the gas group and the 'T' code. For example, an equipment that has Ex ia IIC T6 inscribed on it, the inscription has the following meaning:

- **Ex**: the symbol for explosion protection;
- **ia**: indicates the type of protection utilized for the equipment (intrinsic safety in this case);
- **IIC**: indicates gas group. The Roman II indicates surface industry (and not mining). IIC is the most onerous sub-group in group II such that once an equipment is suitable for IIC it can also be for IIA and IIB;
- **T6**: indicates temperature classification and identifies the maximum temperature ignition-capable surface temperature that the equipment will exhibit. This should be less than the auto-ignition of the hazardous material being processed in the facility.

Internal Combustion Engines

Internal combustion engines (either fueled by gasoline, diesel, natural gas, propane or other fuels) present ignition hazard in petroleum facilities (and indeed in any facility processing flammable liquids and gases). Some examples of such internal combustion engines are compressors, electric generators, automobiles, cranes, forklifts, earthmoving equipment, pumps, fire trucks, and welding machines. An internal combustion engine functions properly with a specific fuel-to-air ratio. When it is located in an atmosphere with flammable gas or vapors, the gases or vapors enter the engine cylinder with the air (and also drawing fuel from its designed supply source) thereby altering the required fuel-to-air ratio. This alteration creates ignition hazards in the following ways:

- Elevating operating temperatures of the engine – Fuel-rich fuel/air mixture causes pre-ignition within the engine. This happens when the fuel-rich mixture gets ignited before the spark plug fires. When there is pre-ignition it creates damaging pressure surges and raises the surface temperatures of the engine and exhaust system. If the surface temperatures reaches the auto-ignition temperature of the flammable gas in the atmosphere, there will be fire or explosion;

- Creating sparks – When a fuel-rich mixture gets in an engine, it may result to incomplete combustion. If this non-combusted gas goes into the hot exhaust system it may ignite, thereby releasing sparks and flames that subsequently ignites the flammable atmosphere;

- Over speeding engines – A fuel-rich mixture can also make engines run faster than designed. When this happens it has the potential to cause catastrophic failure of the engine block leading to fire or explosion.

In-order to control the risk posed by internal combustion engines in hazardous locations, the following preventive measures shall be put in place:

1. As much as possible, machines powered by internal combustion engines should not be installed in or brought into classified locations. If the installation cannot be avoided, the following ignition preventive devices shall be installed in the engines:
   - Intake flame arrestors and exhaust system spark arrestors;
   - Automatic overspeed shutdown devices (ignition kill for gasoline and air shut-off for diesel).

2. Administrative controls, viz:
   - Use of work permit system to control use of internal combustion engines;
   - Adequate information and training of personnel on hazards of internal combustion engines in classified areas and emergency procedures;
   - Establishment of safe traffic routes with adequate warning signs.

Hot Surfaces

Some petroleum processes use direct fired heaters or steam that raise the surface temperature of the equipment. Some examples of such equipment are process heaters, furnaces, dryers, boilers & steam pipes. These hot surfaces are ignition sources and have potential for fire once they come into contact with flammable/explosive atmosphere (and the surface temperature is up to the ignition temperature of the flammable gas). Such equipment shall be identified and handled at the early layout/ spacing stage of the design. To minimize the risk, several measures could be adopted. Some of these measures could be appropriate spacing, insulation, use of high integrity fuel and process pipelines in close proximity to the hot surface, adequate ventilation so that the hazardous gas does not stay long on the hot surface and means of rapid detection of explosive/flammable atmosphere.

Static Electricity

The hazards of static electricity as an ignition source is significant in the petroleum industry. The potential for buildup of static electricity is high and if discharged in a flammable environment it may cause fires and explosions. For this to occur, four conditions shall be present:

- Means of generating an electrostatic charge;
- Means of accumulating a static charge;
- Electrostatic discharge (ESD);
- Flammable mixture.

Generation: Generation of an electrostatic charge takes place by the contact and separation between surfaces of two different materials or objects. A good example is the movement of gases, liquids or solid particles relative to any other material. Typical examples in the petroleum industry are:

- Hydrocarbon flowing through pipes or hose or through filters;
- Splash-filling (if discharge pipe/hose is not covered by the fluid);

- Agitation or bubbling;
- Two phase flow (solid and liquid or liquid and gas);
- Fluids in turbulent contact (e.g. gas or water flowing through liquid hydrocarbon);
- Abrasive blasting;
- Conveyor or drive belt motion.

Accumulation: Accumulation of electrostatic charge occurs when materials are poor conductors or good insulators. This happens when the generated charge does not dissipate because the material is of low conductivity or is insulated from other material. Also when there is not enough time for the electrostatic charge to dissipate. For examples, liquid hydrocarbons (especially refined products) have low conductivity whereas salt water or heavy crude have higher conductivity. Hence, the former has the tendency to accumulate electrostatic charges more than the later.

Electrostatic discharge (ESD): A hazardous electrostatic discharge (ESD) takes place when accumulated charges are released in the form of sparks. If the sparks have sufficient energy to ignite a flammable atmosphere, you have fire and explosion. When charges are generated, there is finite ability of a system to accumulate them and once the ability to accumulate is exceeded, the associated field will cause ionization, leading to the breakdown of the accumulation. In other words, when the voltage of the accumulated charge is high enough to overcome the dielectric strength of air, the air is ionized and forms a conductive path for the charges (just like electricity) to flow. In some cases, sufficient positive and negative charges may accumulate on two different materials. When the attraction becomes strong enough, the charges bridge the air gap between the materials. This bridging heats up the air and more charges jump the gap thereby raising the heat even further. These show up in the form of sparks or glows and is called electrostatic discharge (ESD).

Flammable mixtures: Flammable mixtures are found in classified areas of a petroleum facility. If electrostatic discharge occurs in a classified area and the energy released in the process is equivalent or higher than the ignition energy of the flammable atmosphere, there is the potential for ignition, fire and explosion.

Ignition from electrostatic discharges can be controlling by the following measures:

1. Decreasing static generation – Static charge voltage can be prevented from getting to the sparking potential by decreasing the generation rate. This can be achieved by minimizing the activities that produce static charges. Since static charge is produced when two dissimilar materials are moving relative to one another, the rate of generation can be decreased by slowing down this movement. This can be achieved by, for example avoiding splash-filling of liquids into vessels, reducing flow velocity, decreasing jet and propeller blending;

2. Bonding and grounding - Bonding is electrically connecting two conducting bodies and grounding is when the earth is used as part of the bonding system. Bonding of two bodies stops the accumulation of charges by dissipating them as quickly as they are generated. In other words, bonding prevents difference in potential across the gap between the two bodies, hence no charge will accumulate and no sparks will occur. Grounding is used when a body that is potentially charged is insulated from the earth. The grounding is used to go round this insulation and ensure that the charges do not jump from the body to the earth thereby removing sparks;

3. Use of anti-static additives – The more conductive a material is the better the dissipation of generated charges. There are anti-static additives that could be used in liquids to increase their conductivities and reduce accumulation of electrostatic charges;

4. Increasing ventilation and inerting – If static discharge cannot be controlled by decreasing generation, bonding, grounding, or introduction of anti-static additives, ignition may be prevented by ventilation and inerting. Inerting displaces the air with inert gas and prevents flammable mixture being formed. Mechanical ventilation dilutes the flammable mixture to below lower explosive level thereby preventing ignition.

Lightning Strikes

Lightning strike is a natural form of electrostatic discharge. To control ignition from lightning, the only acceptable method is to dissipate the charge.

1. Metallic structures that are buried or grounded provide safe dissipation of lightning strikes. These include metallic fixed roof tanks and other equipment that are bonded to the ground and buried pipelines.
2. For floating roof tanks, even when grounded, lightning strikes can still ignite gases escaping from the rim seals. The method of control of such ignition is to install straps (shunts) on the circumference of the roof, between the floating and the metallic shoe sliding on the inside of the shell. These shunts will dissipate the charge without igniting gases.
3. For non-metallic structures, protection can be achieved by means of lightning rods, overhead wires or conducting masts.

Hot Work

Hot work simply means work that could produce a source of ignition or ignition sources are being used. Hot work produces sparks, flames or rise in temperatures high enough to start ignition of flammable gases or combustible materials. Some of the activities are welding, cutting, soldering, grinding, abrasive blasting, using of spark-generating tools, use of internal combustion engines etc.

Control of ignition in hot work is done through the work permit system. The system ensures that hot work is not conducted in hazardous areas without taking the required controls to ensure that the ignition sources do not initiate fire or explosion. Some of the requirements of hot work permit system are as follows:

1. Work permit system should be completed and issued before the ignition source is initiated;

2. It is issued by the authority operating the facility or asset where the work is to be done and received by the authority performing the hot work;

3. Flammable products/materials shall be isolated and secured that they do not come in contact with the ignition sources as to cause fire or explosion;

4. Gas testing and monitoring shall be done for presence of flammable gas. Hot work shall not be started or discontinued if gas testing detects flammable gas above 0% of lower explosive limit (LEL);

5. All equipment to be used in the work activity should be inspected to ensure that they are safe to be used;

6. Fire-fighting equipment shall be in place and trained fire watch available;

7. Individuals performing the work activities shall be trained and understand the hazards of hot work in classified environments;

8. It shall specify other necessary steps to minimize/eliminate the risk of ignition and actions to take in case ignition does occur.

CHAPTER 11

EMERGENCY EVACUATION SYSTEMS

In the event that all control measures fail, the next step should be to protect personnel. Every facility should have emergency evacuation, escape and rescue systems. The essence of this is to ensure that all personnel are gotten out of harm's way and taken to safety. The basic principles of an emergency evacuation, escape and rescue system are provision of following:

- Means to communicate to personnel that there is an emergency situation in place (emergency alarm system);
- Channels and means for personnel to leave the hazardous zone and move to a safer place (escape route);
- A safe area or temporary refuge where people can stay till the emergency is brought under control (assembly area or muster point);
- Where the situation deteriorates and the safe area or temporary refuge is threatened, a means to leave the facility completely (evacuation route and rescue system).

Emergency Alarm System

Every petroleum facility should have an emergency alarm system (that is tied to the alarm systems as discussed in Chapter 6). This system is required to have audible and visual means to inform personnel that a life threatening emergency has happened. The alarm should have spoken messages giving clear instructions on the possible level of what has happened and actions required of the personnel.

For this to be more effective, leadership (an effective command and control system) should be in place to take control of the situation as soon as the emergency alarm is given. There should be an emergency response team (under the leadership of an incident commander) that will take control of operations. The incident commander should be the most senior personnel always in the facility and team, selected personnel trained to respond to the types of emergencies likely to occur in such facility. The commander and team should ensure that the instruction of the emergency alarm are adhered to.

Escape Routes

At the design stage of a facility, provision should be made for escape routes from wherever personnel are likely to be present. In every location in the facility there should be two or more escape routes, carefully selected such that if one is blocked (or has been made hazardous by the emergency event), the alternatives could be used. An escape route should have features, as follows:

- It should direct personnel to the assembly area or muster point;
- It should be clearly marked (usually with arrows pointing in the expected direction of movement) with "running man" pictogram and well illuminated in darkness or impaired visibility. The signs should be installed at the average eye level;
- The main escape route should be on the outskirt of the facility, and as straight as possible. As much as possible, routes should minimize difference in elevation. If there are unavoidable change in levels, steps or tripping hazards, these should be well marked;

Assembly Area or Muster Point

Assembly areas or muster points should be carefully selected, away from petroleum process or storage facilities. There should be at least two in a facility, located upwind or cross wind of the prevailing wind direction. As much as possible, the location should be such that there is no process or storage facility between the area and emergency evacuation route. In offshore facilities, it should preferably be in the living quarters and closer to the helipad than the process units. In situation of massive gas release, the assembly or muster area could be indoors. Such indoor assembly areas should be buildings specially designed to withstand explosions and ingress of gas.

At the assembly area, a designated member of the emergency response team (ERT) should conduct a headcount and report to the incident commander. There should be a system in place to determine the number and identities of people inside a facility at every point in time. With a headcount, missing persons can be identified and decision made by the incident commander to launch search and rescue. When all the people are accounted for and/or the assembly area is threatened by the emergency event, the ERT member responsible for the safe area can organize the evacuation of the people.

Evacuation

Evacuation is a planned method of safely taking people from a facility to a safe location. The evacuation route should be linked directly to the assembly area or muster point without passing through the hazardous zone. This should be determined at the design stage and built into the facility. In offshore or swamp installations, an evacuation involves having means of transporting people to land. This is usually achieved by first moving them quickly to safety using lifeboats, life rafts or totally enclosed motor-propelled survival craft (TEMPSC) and then using service boats to rescue them. Helicopters can also be a means of quickly transporting people to safe locations but in some emergencies it might not be safe for a helicopter to approach.

Testing and Inspection

For emergency evacuation, escape and rescue systems to work as designed when required, they have to be periodically inspected and tested. In addition to periodic tests and inspections, emergency response drills should also be opportunities for checking how the various systems work. During a critique meeting that follows drills, findings should be discussed and action plans put in place to address observed lapses.

CHAPTER 12

FIRE SUPPRESSION SYSTEMS

There are few fires in petroleum process facilities that self-extinguish (that is, without intervention). There are various suppression systems that can be deployed to ensure that when fire occurs it does not burn out of control, viz:

1. Water-based Extinguishing Systems;
2. Halon Extinguishing Systems;
3. Carbon Dioxide Extinguishing Systems;
4. Chemicals Extinguishing Systems;
5. Foam Extinguishing Systems.

Water-based Extinguishing Systems

Water and water-based systems are the most common and widely used fire extinguishing systems in the petroleum industry. Water as an extinguishing agent is very effective due to its properties, which makes it able to cool, dilute or emulsify the fuel and remove oxygen.

However, it is not suitable for all classes of fires. There are five typical classes of fire: (1) class A – solid combustibles; (2) class B – flammable and combustible liquids/gases; (3) class C – electricals ; (4) class D – combustible metals; (5) class K – cooking oils. Aside class A, the choice of water for other classes of fire must be weighed carefully against inherent hazards on people. Some of these hazards are electrocution in electrical fires and adverse reactions in liquids and combustible metals. In all classes, there is the potential hazard of inhalation of or exposure to high temperature steam.

There are some compounds that when added into water, change some of the properties and make them suitable for use in several classes of fire. Some of these additives are: wetting agents, these tend to change the surface tension or viscosity and make water more efficient in quantity and timing; foaming agents, tend to blanket flammable/combustible liquids and smother/remove oxygen. In gas fires, while attempts are being made to isolate the fuel source, water can be used to cool nearby equipment and the one on fire.

A typical water firefighting system consists of supply/storage, pumping, distribution and delivery units. A supply/storage unit could be storage tanks, firewater lagoon, or natural water bodies (like lake, river or sea). Pumping units consist of pumps and prime movers that move sufficient quantity of water at the right pressure from supply/storage point to fire incident locations. The distribution system consists of network of pipes in the form of ring main through which the water is moved by the pumps to the fire. Hydrants, monitors, sprinklers and other devices are connected to the ring main and are the means by which the water is delivered to the point where the fire is extinguished.

Hydrants are above ground pillar-like connection points to the ring main from which fire fighters tap into the fire water supply system. A hose is attached to the fire hydrant and on opening a valve on the hydrant, the water flows through the hose to where the user directs it. The other end of the hose can also be connected to a fire truck, which uses its pump to boost the water pressure and can split it into multiple streams.

Figure 15: Fire Water Hydrant

Monitors are fast-acting fire water delivery systems used primarily to protect high risk areas from quick fire spread. It consists of a valve and control system. It can also be adapted to mix foam into the water supply for fighting pool fires.

Sprinklers are active fire protection devices that work with little or no human intervention. They are usually activated by smoke or heat detection systems. There are various types of sprinkler systems. A pre-action type is filled with air and when activated, allows water to flow through and sprinkles on the equipment being protected. A dry pipe sprinkler is also filled with air but on activation, the air spouts before water escapes. Wet pipe sprinklers are always filled with water and react quicker to fire than the two previous ones.

A deluge fire protection system is also a type of sprinkler. The sprinkler heads are open and the pipe that connects them to the fire water system is not pressurized. It has a deluge valve that opens on activation (by smoke or heat detection system) and allows water to flow to the sprinkler heads.

Fire water systems can either be dual-type, servicing both fire protection and domestic/utilities, or separate, only for fire protection. In the case of dual systems, some of the pumps (known as jockey pumps) are continuously in operation to ensure that water is always available at the right pressure. When there is heavier demand for water than the jockey pump can cope with, the distribution pressure will drop. The system should be designed in such a way that a certain drop in pressure triggers other pumps to start automatically to compensate for the higher demand. An advantage of the dual-type is that faults may be noticed earlier before a fire incident.

The advantages of separate type are: the stored water is dedicated sorely for fire protection and hence little risk of non-availability when there is fire; the system is designed well to meet fire demands; fire protection personnel have absolute control; and cost of water treatment to portable quality is eliminated. A disadvantage is that the pumps may not be running continuously and are required to be started when there is demand. If proper inspection, testing and maintenance is not conducted regularly, they may fail to function as required when they are needed.

Figure 16: Fire Water Pump House

Halon Extinguishing Systems

Halon is a liquefied compressed gas and it is a clean fire extinguishing agent. It halts the spread of fire by chemically disrupting combustion. It is referred to as a clean agent because it is electrically non-conducting, volatile or gaseous fire extinguisher that leaves no residue on evaporation. It is effective for classes A, B & C fires.

For fire to start or continue there has to be a combination of three things, viz; fuel, oxygen and ignition source. This combination is called the fire triangle. Traditionally, in-order to extinguish a fire one arm of the triangle has to be removed. In the case of halon, it stops a fire by breaking the chain reaction. It chemically reacts with each of the three and stops them from reacting with each other.

Halon is useful in protecting sensitive installations like computer, communication equipment and turbines because it has the ability to put out fire without leaving residues that may further damage the equipment being protected.

However, halon is a CFC (chlorofluorocarbon) that are harmful to the ozone layer and contribute to climate change. The production of new halon was stopped in 1994 under the Clean Air Act but recycled halon fire extinguishers can still be purchased and used. It is of low-toxicity, chemically stable and relatively safe for human exposure.

Carbon Dioxide Extinguishing Systems

Going back to the fire triangle, carbon dioxide as an extinguishing agent works by eliminating the oxygen to stop a fire. It is triggered by fire or smoke detection systems installed in the equipment it is protecting. When triggered, the system quickly releases carbon dioxide gas into the space/equipment it is protecting. As the level of the gas increases the oxygen decreases quickly thereby making the fire to be extinguished or prevented from developing.

Carbon dioxide gas is odorless, colorless, electrically not conductive and leaves no residue after acting. It is therefore very useful for protecting sensitive equipment like computer, electronic devices, turbines, power stations, engine rooms, flammable liquid storage rooms and other industrial machines.

However, it poses serious health risk to humans. Hence it can only be deployed in spaces that are not normally manned. Most carbon dioxide extinguishing systems are normally designed to have about 35% concentration and a concentration of even 7.5% can lead to asphyxiation on human exposure. Protected spaces are required to have safety devices to ensure that people are not exposed. These include visual and high-sounding audio alarms that warn people prior to release. The period between the alarms and the actual release is set to be enough for people to safely evacuate. There should also be safety signs and adequate training of personnel who work in such environments.

Chemical Fire Extinguishing Systems

Chemical fire extinguishing agents are either dry or wet. They are made of powders, and the chemical contents are commonly sodium bicarbonate and mono-ammonium phosphate. Sodium bicarbonate extinguishes classes BC fires and mono-ammonium phosphate, classes ABC. A typical chemical extinguishing system consist of pressurized tank filled with the dry powder and high-pressure nitrogen cartridge. When there is fire and the system is activated, the high-pressure cartridge will discharge and open the valve of the powder tank, releasing the extinguishing agent into protected space/equipment to stop the fire. The system is recharged in-order to be prepared to work again. These types of extinguishing systems are efficient, easily accessible and non-conductive. They are made for ordinary combustible materials, flammable liquids and electrical equipment. However, dry chemical agents leave residue after putting out the fire and need to be cleaned.

Wet chemical agents are made of either potassium acetate, potassium carbonate or potassium citrate. They are used to extinguish burning liquids like cooking oil and act by reacting with the hot oil, forming a soapy foam blanket over the fire and stopping the fire. They are useful in extinguishing kitchen fires.

Foam Extinguishing Systems

Foam fire suppression agent is a mixture of water, chemical compounds and air-filled bubbles. When released, it floats on the surface of combustible liquids to stop vapor formation, separate the fuel from the air and the water content cools the fuel to extinguish the fire. There are fixed foam monitors that are used to fight fire from a safe distance. There are also portable applicators and nozzles used to get to areas where the fixed system cannot. Provision for foam systems should be made where there is bulk storage of liquid hydrocarbons.

Figure 17: Foam System

Portable Fire Extinguishers

Portable and wheeled fire extinguishers should also be available in different areas of a petroleum facility for use in suppressing fires at the incipient stage and along emergency escape routes. The types, sizes and locations of these extinguishers will depend on facility configuration and potential fire hazards.

The common types and their various uses in a typical petroleum processing facility are as follows:

Dry Chemical – this type is mainly for classes A, B & C fires and the color of the tank is typically red with white band;

Dry Powder – this type is for class D fires and colored red with blue label;

Carbon Dioxide – the typical color of this is black (or red with black band) and meant for classes B & C and typically in areas where residues may create more problems. It is more effective in enclosed spaces because carbon dioxide acts by displacing available oxygen;

Foam – this is for classes A &B. The color code is cream. It is either the tank is completely colored cream or red with cream band.

Inspection, Testing and Maintenance

Fire extinguishing systems and equipment should be tested on installation by competent people or organizations. They should be inspected regularly by the users (at least monthly) for defects. Competent persons should be invited to effect repairs, as required, when defects are observed. At least, annually, they should be inspected and/or tested by the competent people. Records of all inspections, tests and maintenance activities performed on them should be kept. Specific inspection, testing and maintenance requirements for the different types of fire extinguishing systems can be found in:

- NFPA 10, Standard for Portable Fire Extinguishers;
- NFPA 12, Standard on Carbon Dioxide Extinguishing Systems;
- NFPA 12A, Standard on Halon 1301 Fire Extinguishing Systems;
- NFPA 17, Standard for Dry Chemical Extinguishing Systems;

- NFPA 17A, Standard for Wet Chemical Extinguishing Systems;
- NFPA 25, Standard for the Inspection, Testing and Maintenance of Water-based Fire Protection Systems;
- NFPA 291, Recommended Practice for Fire Flow Testing and Marking of Hydrants;
- NFPA 2001, Standard on Clean Agent Fire Extinguishing Systems.

CHAPTER 13

OPERATING PROCEDURES

For every petroleum process facility, the owner/employer should develop and implement written operating procedures. The procedures should give clear instructions on the steps to be taken or followed in running the plant or equipment. The process safety information package is a good resource to ensure that the procedures are consistent with the hazards of the chemicals in the process and the operating parameters are accurate.

The operating procedures should include operating parameters like pressure limits, temperature ranges, flow rates, level limits, alarms, and actions to take during upset conditions. It should indicate distinction between startup, normal operation and shutdown.

A procedure should describe tasks to be performed, data to be recorded, samples to be collected, operating conditions to be maintained, and safety precautions to be taken during startup, normal operations and shutdown. It should be technically correct, effectively communicated to employees and periodically revised by engineering and operations personnel to ensure it reflects current realities.

It has to contain the following details, as a minimum:

1. Steps for the following operating phases: initial startup, normal operations, emergency shutdown, emergency operations, normal operations, normal shutdowns, and startup following maintenance;

2. Operating limits: temperature, pressure, flow rate, level, alarms, consequences of deviation from limits, and steps to correct deviations when they occur;

3. Safety and health: properties and hazards of chemicals in use, control measures to prevent personnel exposure and recovery measures, safety systems and their functions;

4. In case of computerized process control systems, the procedures should describe the software logic and the relationship between the equipment and the control system.

It is important that the procedures be well communicated to personnel, readily available for use and appropriate training provided, where necessary. In situations where workers are not fluent in English, it may be considered to prepare the procedures in a second language.

Whenever there is a change in the process, technology, equipment or chemicals, the procedures should be reviewed and communicated to the employees that work with them. During the reviews, employees should be consulted to ensure that practical lessons are captured. On annual basis, the facility owner should certify that the operating procedures remain current and accurate.

CHAPTER 14

EMPLOYEE TRAINING

It is imperative that all employees (operations, maintenance, engineering and contractors) that are assigned in a petroleum process facility be adequately trained. The training program should include the safety and health hazards of all chemicals and the processes the employees work with so that they can protect themselves, fellow employees and the public. Operating procedures, safe work practices, work authorization (routine or non-routine), emergency response, and other safety procedures should also be part of the program.

The facility owners or managers should perform training needs analysis that identifies the employees to be trained, the specific subjects for each employee to cover and the objectives. The objectives should be written in clearly measureable terms and the required actions to demonstrate that these objectives have been met.

The training should be conducted not only in the class rooms but also include hands-on sessions so that employees can apply what they have been taught. As much as possible, the hands-on sessions should have simulated or real situations.

In addition to the initial training, refreshers should be provided at least every three years or whenever there are changes in chemicals, process, equipment or technology. In situations where there are changes, the employees should be taught the effects of the changes on their job tasks. Adequate records of all these should be kept.

Periodically, the training program should be evaluated to check that the necessary skills, knowledge and routines are understood and implemented by the employees that receive the training. If the result of the evaluation shows that the level of desired skill and knowledge is not being achieved, the program can be reviewed in consultation with the trainers and trainees.

CHAPTER 15

NON-ROUTINE WORK AUTHORISATION

Work activities that are non-routine or potentially hazardous are planned, authorized and carried out using work permit system. A work permit system ensures that an activity has been properly planned, consideration given to the risks involved and that all concerned parties have been communicated. When a permit is issued, it authorizes certain people to carry out specific work, at a certain time, at a specific facility, using specified equipment and sets out the main precautions required to complete the work safely.

Each organization or company should develop its own work permit procedure but the essential features of the system are:

- Identification of types of work considered as non-routine and potentially hazardous, and facilities applicable;
- Clear roles and responsibilities (identifying who may authorize specific types of jobs, who is responsible for specifying precautions, who is responsible for checking the work site before and after, who is responsible for ensuring that specified precautions are followed);
- Training in the issue, use and closure of work permits;
- Monitoring and auditing to ensure that the system is being implemented as required.

Work permits should be considered whenever:

- The proposed job is 'non-production' (for examples, maintenance, repair, inspection, testing, construction, dismantling, assembling, cleaning, modification, adaption);
- The activities are in a facility specified in the organization's work permit procedure, non-routine and require some kind of job safety analysis;
- Two or more different set of crews or different contractors are in a site or facility conducting different jobs and need to co-ordinate their activities to ensure that their work is completed safely.

It should however not be used for activities considered as low risk so as not to trivialize it thereby weakening the overall effectiveness.

Responsibilities

In an organization's work permit system, the following individuals should have specific responsibilities defined in the work permit procedure.

1. Employers or duty holder or installation owner or company chief executive should ensure:
   - An appropriate work permit system is developed;
   - Adequate resources to implement the system is provided;
   - Training programs and competence standards to implement the work permit system are established and maintained;
   - Monitoring, auditing and review of the system are established and maintained.

2. The facility manager should ensure that:
   - All work and facilities requiring work permit are identified and specified;
   - Personnel who have roles in the administration of work permit have the competence specified in the procedure;
   - The administration of work permit in his area of responsibility is properly coordinated;

- The system is monitored and audited to ensure effective implementation.

3. Contractor's and sub-contractor's management should ensure that:
   - They are aware of and understand the work permit procedure for the locations where their employees are to work;
   - Their employees have the competence specified in the procedure, understand the operation of the system and their specified responsibilities within it;
   - They have adequate resources to implement the system.

4. Operations Supervisors should ensure that:
   - As permit issuers for their units, that they have sufficient knowledge about the hazards associated with the facility and be able to specify control measures;
   - The nature of the work is clearly specified and understood;
   - All associated hazards are correctly identified and necessary precautions are in place before commencement of work (including isolation of energy sources and gas testing);
   - The permit receiver (or person in charge of the work) has the competence specified in the procedure to fulfill his responsibilities;
   - Everybody who may be affected by the activity is informed and kept in the communication loop until the work is completed;
   - There is a joint site inspection prior to issuing and close-out of the permit;
   - Site is monitored throughout the duration of the work to ensure that the necessary precautions remain in place or otherwise withdraw the permit and stop the job;
   - If the work extends beyond his shift, that there is effective handover to the supervisor of the next shift;
   - Copies of issued work permits are kept for the period specified by local legislation.

5. Permit receiver (or person in charge of the work), whether employed by the duty holder or contractor, should ensure that:
- He and his crew have the competence specified in the work permit procedure to fulfill their respective responsibilities;
- The scope of the job is discussed fully with the permit issuer;
- Joint inspection of the site is conducted with the permit issuer to ensure that all necessary precautions are in place (including isolation of energy sources and gas testing);
- His crew is briefed on the details of the permit, potential hazards and necessary precautions;
- All precautions are in place throughout the duration of the work;
- On completion or suspension of the work, the facility is made safe and joint site inspection conducted before closure of the permit and hand back to the issuer.

Work Permit Types and Forms

The center of the work permit system is the paper or electronic form that is used in facilitating communication between all the people involved. Every company or organization that needs to operate the system should design its own forms for the different types of work permits and every effort should be made to keep it simple and user friendly.

The essential contents of the form should be the following:

- Permit type;
- Permit number;
- Identity of the facility/plant/equipment where the work will be done;
- Exact work location;
- Description of work to be done;
- Details of equipment and tools to be used for the work;
- Details of potential hazards;
- Precautions necessary and actions in the event of emergency;
- Details of protective equipment to be used;

- Required gas tests, frequency and records;
- Other persons to be notified or countersign;
- Date of work and period of validity;
- Signature of permit issuer;
- Signature of permit receiver;
- Signature of gas tester;
- Signature of handover of responsibilities between shifts;
- Extension of permit after the validity period;
- Signature of issuer and receiver to close out permit.

The forms should be colored-coded as to the types of permits so as to be able to differentiate one from the other. The following are the suggested types of work permits and color coding of the forms:

- <u>Hot Work Permit</u> – hot work permit should be used when the type of work to be done can be a source of ignition, the equipment to be used or the work activity itself can generate sparks, flames, and heat that could ignite materials. Examples are welding, flame cutting, use of grinders, abrasive blasting, use of internal combustion engines. Hot work permit forms could be colored red;
- <u>Cold Work Permit</u> – cold work permit should be used when the work will not produce sufficient energy to cause an ignition. Examples could be masonry work, excavation with hand tools. The color could be blue;
- <u>Confined Space Entry Permit</u> – confined space entry permit could be colored green and should be used when the work involves entry into a place that is considered a 'confined space';
- <u>Breakage of Containment Permit</u> – breakage of containment permit should be used when the work involves breaking open a vessel, equipment or pipeline that has the potential to contain flammable, explosive, hazardous or high pressure fluid. The form could be colored yellow;
- <u>High Voltage Electrical Work Permit</u> – high voltage electrical work permit should be used when the work involves high voltage

electrical equipment or conductors. The form could be colored orange.

Whether the forms are in hard copy or electronic based, it is very essential that the use of each type is understood by everyone involved in the activity.

Competence and Training

A work permit system is only as good as the competence of the people who operate it. The employer's chief executive (or duty holder), facility manager, contractors & sub-contractors management, permit issuers and receivers should have sufficient knowledge of the system.

In addition, work permit issuers/receivers should be in supervisory positions and should be competent in their respective professions or trade. The issuer should have sufficient knowledge of the facility in which he is working, viz: plant and equipment layout; the process (production or drilling); potential hazards existing; means to control the hazards in order to make the facility safe for the work to be done; required actions in case of emergency; and company safety rules/procedures.

Both the issuer and receiver should attend training on the requirements of the work permit system, which should include:

- Local legislation and company policy on work permit system;
- Industry guidance and case histories of incidents whose root causes are failure of work permit systems;
- Gas testing and monitoring;
- Basics of hazard recognition and job safety analysis;
- Safety requirements for work in confined spaces;
- Isolation procedures, lock out and use of hold tag;
- Employer's work permit procedure or system.

At the end of the training, a written examination should be conducted to assess the level of understanding of the candidates. Those who are successful in the examination should be given certificates and be designated as 'Competent Persons' to issue and receive certificates with respect to their respective areas of trade or profession.

Issuance and Use of Work Permits

When work is to be carried out in facilities where work permit system is applicable:

- The permit receiver (responsible for carrying out the job) and his team should conduct a job safety analysis of the proposed activity;
- The receiver should request a work permit from the issuer, specifying the scope of work and duration (and submitting a copy of the job safety analysis);
- The issuer should go through the scope of work, the job safety analysis and determine the correct type of permit for the proposed work (it is normal for one activity to require two or more types of permits);
- The issuer should verify that the receiver has a valid work permit certification for the type of work to be performed;
- The issuer and receiver should jointly complete the work permit form, specifying precautions as in the job safety analysis (the issuer can put extra precautions with his knowledge of the facility);
- The issuer and receiver should conduct a joint inspection of the site where the work will be done in order to: take note of the state of the facility and possibly any hazards not already identified; conduct gas tests; put isolations in place (if required); and confirm that the facility is safe for the work to start;
- The issuer and receiver should then sign the permits and if a counter-signature is required, obtain such;
- When all relevant signatures have been gotten, the work can start;
- The permit is signed in duplicate (as a minimum), the original being with the issuer and receiver taking the duplicate;
- While work is on-going the receiver's copy should be displayed on site and the issuer's copy displayed on the permit board in the office or control room;

- Both the issuer and receiver should be responsible to ensure that the precautions specified in the permit are in place throughout the duration of the work;
- At the end of the validity period, the permit may be extended or closed and a new one issued (depending on the requirements of the company's work permit procedure);
- In a case of shift change, the out-going and in-coming shift permit receivers should conduct joint site inspections, agree that precautions are adequate and sign in designated columns (of both copies of the permit);
- Before permit is closed, both the issuer and receiver should conduct joint site inspection and sign-off that the site has been made safe.

Monitoring and Audit of the Work Permit System

Work permit system should be monitored continuously to ensure that the permits are correctly issued; precautions specified on them are being complied with and properly closed. Discrepancies should be noted and discussed with all concerned parties.

Implementation of the system should be audited at least annually and findings used to ensure continuous improvement.

CHAPTER 16

ISOLATION OF PLANT & EQUIPMENT

Process plant and equipment periodically undergo intrusive maintenance, inspection and sampling. These activities have the potential of releasing hazardous substances with consequences like fire, exposure of people to high pressure and hazardous materials, severe injuries and fatalities. In this chapter we will discuss on how to manage the isolation of process plant and equipment in-order to perform such activities safely. Process equipment in a plant are connected to one another by series of in-plot pipes. Some of the maintenance, inspection and sampling jobs are conducted in an equipment while the rest of the plant is live. Isolating the equipment from the rest of the plant simply involves closing connecting piping to prevent ingress of hazardous substance from the live portion to the isolation envelope. The same basic principle is applicable when a plant is shut down for maintenance (like during turnaround and inspection). In this case the entire plant becomes the isolation envelope. The isolation of the plant would be achieved by first closing the piping (at battery limits) connecting the envelope to other upstream or downstream facilities or utilities. After that, the individual equipment can be isolated depending on the scope of work.

These isolations are achieved using valves, blinds, spades and blank flanges. Below are simple definitions of these isolation devices.

Block Valve – A valve that controls passage (permitting or stopping flow) of a fluid through a pipe.

Blind/Blank flange – A solid flange that is bolted to a flange connection on pipes to provide positive isolation (completely block flow) between sections of a process.

Spade – A circular flat plate (normally with a handle) that is bolted between two flanges to block flow. A spade (unlike blind and blank flange) does not have bolt holes in it. It is sometimes referred to as slip blind.

Spectacle blind – Consists of two circular plates connected by a straight handle, with one section open and the other closed. The closed section is designed to block flow and the other to allow flow. It looks like a pair of spectacles (or the number 8), hence the name "spectacle blind". It is bolted between flanges on piping and equipment.

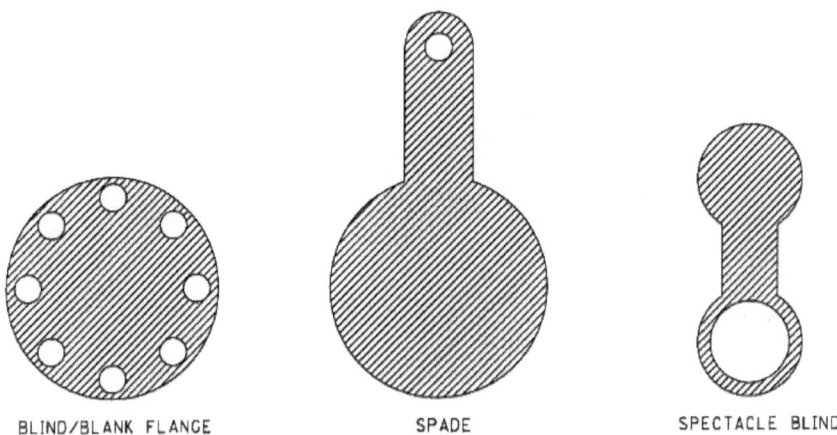

BLIND/BLANK FLANGE   SPADE   SPECTACLE BLIND

Figure 18: Types of Blinds (blind/blank, spade and spectacle)

Vent or drain valve – A valve on a smaller bore piping attached to the main piping that can be opened to release pressure and confirm whether or not there is pressure on the downstream of a block valve.

There are four regular methods of applying these (singularly or in combination) to ensure safe isolation of a plant or equipment. The method chosen depends on the type of maintenance to conduct, the nature of the hazards of the material (process or utility), the nature of the plant or equipment and the environment. The uses of each of these methods are discussed below. There are also several other specialist isolation techniques that are beyond the scope of this book.

Single Block Valve Isolation

This is done by closing a block valve on the piping. It is the fastest, simplest and cheapest method of isolation and used for all fluids at various pressures. It does not require intrusion into the pipe and hence no significant hazard to personnel conducting the isolation. This does not require specialist training.

This methods has a disadvantage that the valve may be passing (not shutting-off tightly) due to damage of the seal. It is important to have a vent/drain valve or pressure indicator downstream of a single block valve in-order to ascertain that it is not passing before maintenance work is allowed to start. The device has to be secured from being inadvertently opened while work is in progress by locking it and tagged to properly identify it.

This method of isolation is used when quick routine maintenance jobs are to done (for examples, changing valves, filters or gauges). It should never be used when hot work, entry into confined space or activities that will last a long duration (say days) are involved.

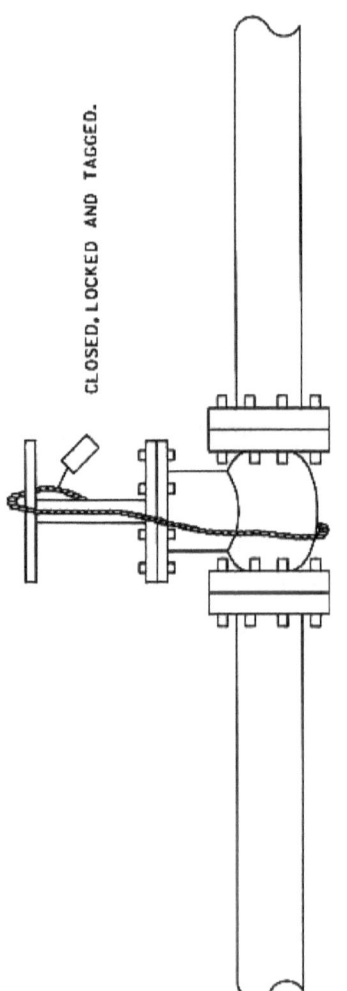

Figure 19: Single Block Valve Isolation

Double Block & Bleed Isolation

This is a combination of two block valves in series and a bleed (vent or drain) system in between. If the fluid is sour, flammable or in any way constitutes hazard to people, the bleed system should be connected to a closed drain or flare system. Double block and bleed is recommended for isolation in all sour systems or in fluid rated above 900#. This method is more secure than single block valve but there is still the potential for the bleed system to be plugged and hence still not recommended for hot work and confined space entry activities. The two block valves must be locked in closed positions and the vent valve locked in open position before work is permitted to start.

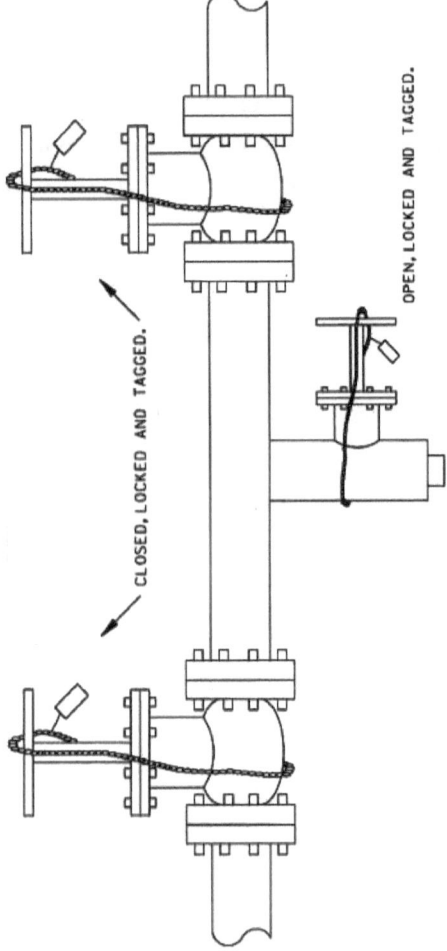

CLOSED, LOCKED AND TAGGED.

OPEN, LOCKED AND TAGGED.

Figure 20: Double Block and Bleed Isolation

Blinding

This isolation method involves installation of a blind between two flanges to completely separate the plant/equipment (positive isolation) to be worked on from other parts of the system. Valve isolation (either single block valve or double block and bleed) is a requirement during blind installation. After the valve isolation, it must be proven to be holding (not passing) before the flanges are separated to insert the blind. Blinding is suitable for all fluids over a range of pressure ratings. It is a cheap method of isolation, does not require specialist training but relatively slow to install and there is risk of exposure of installation personnel to hazardous fluid.

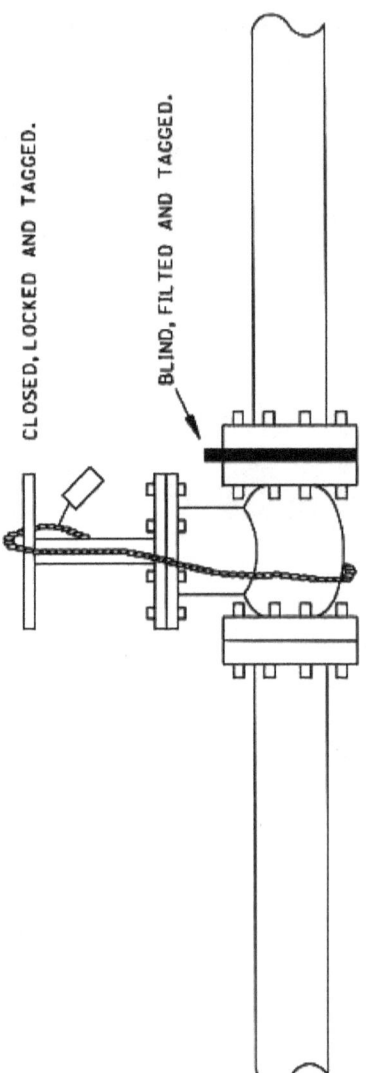

CLOSED, LOCKED AND TAGGED.

BLIND, FILTED AND TAGGED.

Figure 21: Blinding Isolation

Pipe Disconnection

Pipe disconnection is a combination of valve isolation, removal of pipe spool and installation of blank flange from where the spool is removed (on the section that contains the hazardous fluid). The blank flange can have drain/vent valve incorporated for bleeding/venting and proving purposes. This method is also suitable for all fluids over a range of pressure ratings and best for isolation for extended period of time. It does not require specialist training but relatively slow and has the risk of exposure of installation personnel to hazardous fluid.

Other specialist isolation techniques are: squeezing off (squeezing polyethylene pipe together with clamp to stop flow); foam bagging (injection of foam into semi-porous bag inserted into a pipe); hot tap and stopple installation; pigging; pipe freezing; inflatable bags and pipe freezing. These techniques require specialist training and equipment in order to conduct the isolation safely and effectively.

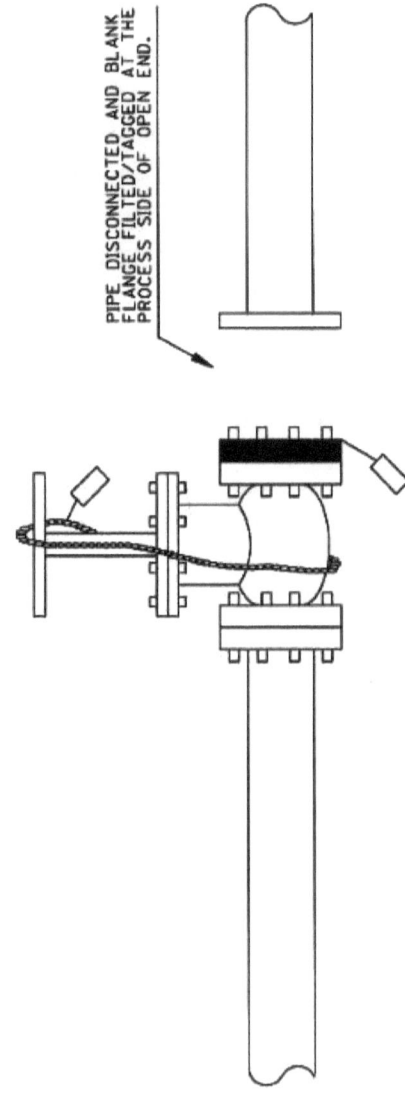

PIPE DISCONNECTED AND BLANK
FLANGE FILTED/TAGGED AT THE
PROCESS SIDE OF OPEN END.

Figure 22: Pipe Disconnection Isolation

Management of Plant/Equipment Isolation at Design Phase

Safe and effective isolation of plant and equipment starts with good design (both for new plants and modifications). A documented philosophy that specifies the strategy for maintenance should be available to guide design engineers. Such a document should give ideas on how and where isolation is required in-order to perform intrusive maintenance and inspection activities. The following should be put into consideration in deciding on the required isolation facilities and ensuring they are incorporated at the design stage to make the plant inherently safer.

- **Positive Isolation Requirements**

Provision should be made for positive isolation at the battery limits of the process, utility, relief and blowdown lines. Blinding devices should also be provided for all inlet and outlet lines into vessels or equipment that entry is required for inspection, maintenance and testing. In addition, where segregation of fluids is imperative or segregation of parts of a plant that might be exposed to overpressure (in alternative operating modes), blinds are required.

In any of these situations where blinding is done regularly, spectacle blind is preferred. Spades and blank flanges are advised to be provided where maintenance is not done regularly. For large pipe sizes, to prevent unnecessary weight on flange connections, spades or blank flanges can be installed instead of spectacle blinds.

The rating of a blind shall be for full process pressure on one side and atmospheric pressure on the other side.

- **Location of Isolation, Verifying and Monitoring Devices**

For every blind, there shall be at least block valve (to conduct primary isolation), and a vent/drain or pressure indicator (to verify integrity of the valve). The verification device (vent, drain or pressure indicator) is required to be downstream of the block valve and preferably between the valve and the blind in-order to also be useful during de-blinding.

As much as possible, blinds locations should be on horizontal lines. This makes installation much easier and reduces the potential of gaskets damage. In addition, blinds installed on vertical lines has the potential of trapping liquids above them. This becomes even worse in vertical steam or water lines with the potential for freezing.

- **Pipework and Support**

The layout of pipework should be such as to allow easy draining of fluid and minimize trapped inventories. The design engineers should ensure that the piping is of sufficient size and appropriate configuration to reduce the probability of blockage in service. If it is envisaged that a spool piece could be disconnected during isolation, it should be easily removable.

Appropriate pipe supports should be installed in areas of pipework where physical disconnection of spools and installation/removal of blinds are required. This is to maintain piping alignment when these activities are undertaken.

- **Pressure Safety Valves**

Block valves should be installed up and downstream of pressure safety valves (PSV). The upstream block valve is required when a PSV is to be replaced while the equipment is in service and there is a spare PSV. The downstream block valve is required if the PSV is discharging to a common header and it is required to be isolated. It is noteworthy that between each of the block valves and PSV, there should be vent or drain to be used in verifying the integrity of the isolation valves.

- **Plant Labeling and Identification**

There should be a scheme to permanently label and identify all process plant, piping, valves, drains, vents and blinds. This scheme should be reflected in the piping and instruments drawing (P&ID) for ease of cross-referencing. This is to eliminate or minimize mistakes in identification during isolation activities.

- **Access, Handling and Lighting**

Suitable access and adequate lighting should be provided at locations of isolation devices. Enough space should be created for ease of unbolting, bolting and swinging of blinds. If cranes are to be used, adequate space should be provided for the positioning of the lifting equipment and slinging of the load. For blinds that are removed on regular basis, platforms should be provided. In addition, for blinds weighing more than 45kg (100lb), it is recommended to provide permanent handling equipment.

Management of Plant/Equipment Isolation at Operation/ Maintenance Phase

Risk Assessment

During the operations phase of a plant, management of isolation starts with a risk assessment of loss of containment of the fluid/substance proposed to be isolated. The inherent hazards should be identified and the consequences of loss of containment during the isolation activity or in the event of failure of isolation. The two classes of hazards to be considered are hazards related to the isolated substance and those associated with the work task.

In identifying the hazards related to the isolated substance the following should be considered: pressure, temperature, chemical composition, quantity of substance that can potentially be released, the population at risk (proximity of the population to the equipment and the speed with which they can be affected) and the escalation potential (in the event of loss of containment).

Hazards related to the isolation/de-isolation work tasks and the work tasks to be conducted in the isolated equipment should be given consideration. For example, if the substance is flammable, hot work shall be identified as a potential hazard. If the work tasks involves entry into confined space, hazards to be considered are flammable or toxic vapors, asphyxiation, oxygen depletion, carbon dioxide build-up and drowning by ingress of fluid.

For identified hazards, the potential consequences to people, the environment and plant/equipment in the event of isolation failures and the likelihood of failure of each type of isolation task shall be assessed. This risk assessment should cover every stage of the isolation activity, viz: preparatory work (like depressurization, draining, venting, purging, steaming, washing out), installation of isolation devices, proving/ monitoring of isolation, integrity of the isolation during intrusive maintenance work and reinstatement of plant after maintenance. For every hazard, adequate mitigation measures shall be identified to ensure the risk is brought down to as low as reasonably practicable (ALARP).

Selection Of Isolation Methods

Earlier in the chapter we have discussed the different methods of isolation, viz single block valve (sbv), double block and bleed (dbb), blinding, disconnection and other specialized methods. These have different levels of integrity. Based on the result of the risk assessment, an isolation method shall be identified for each desired isolation point. The higher the severity of the risk, the higher is the desired level of integrity. Some of the methods (like sbv and dbb) can be used as initial isolation before a final isolation is applied.

Blinding or disconnection should be used for confined space entry, hazardous substances, extended isolation or when failure of isolation could lead to catastrophic consequences.

Isolation Plan/Procedure

The level of planning for process isolation depends on the risk severity, complexity of the equipment/plant and number of required isolation points. Good practice requires that when there are more than two isolation points, a written procedure shall be developed, and adequately reviewed by cross functional team (of operations, maintenance and engineering personnel) to ensure accuracy and ease of implementation.

The procedure should be a step-by-step method of shutting down the plant or equipment, depressurizing, draining, venting, purging, flushing/washing, steaming out and disposal of hazardous substances. It should also specify isolation sequence (e.g. order of valves closures, blinds installation or pipes disconnection). In order to avoid mistakes, the procedure is required to have diagrams (e.g. process and instruments drawings, isometrics, accurate sketches) on which shall be marked valves, pressure indicators, drains, vents, blinds, and disconnection points required for isolation. The procedure and attached diagrams shall be easily cross-referenced.

Before a procedure is approved, it is also important to conduct a 'walk-the-plant' verification to check that the diagram in use matches the installation and the isolation points can easily be identified, accessed and operated.

Installation of the isolation

There are two stages of installing isolations. The first stage is called initial (or primary) isolation and the second, final (or secondary) isolation. The initial isolation usually involves closing upstream isolation valves. Then the downstream section is depressurized and purged. The valve should then be proven to be holding tight (or not passing) by either opening a downstream drain or vent or checking that a downstream pressure indicator is reading zero. When it has been ascertained that the isolation is effective and that there is no hazardous substance, the line can be broken into to install the final (or secondary) isolation.

Before breaking open the line to install the final isolation, precautions shall be taken to ensure that installation personnel will not exposed to hazardous substances. Some of those precautions are as follows:

- Detailed planning of the work – All the blinds, spacers, gaskets, tools, equipment and personal protective equipment required shall be available on site and ascertained to be in good condition. The work procedure (such as gas testing requirements, loosening of first bolts at the "5 o'clock" position away from personnel and other step-by-step procedure), and job safety analysis shall be communicated to the work crew via tool box meeting;
- Restriction of incompatible work– If, for example, the substance in the equipment being isolated is flammable, no hot work shall be allowed within 24 meters horizontal radius of the location;

- Restriction of access – Access shall be restricted to only essential personnel within 24 meters horizontal radius. It is advisable to barricade the area and install appropriate warning signs;
- Personal protective equipment – Adequate number of personal protective equipment, including respiratory protection, shall be available and all relevant members of the crew ascertained to have been trained in the use.

The final isolation, which could either be blinding or pipe disconnection, should protect personnel carrying out intrusive maintenance activities from being exposed to hazardous substance from the plant during the duration of the work. The isolation devices should be secured to prevent them from being tampered with. Security of isolation devices can be achieved by various means, viz: padlocking of valves; removal of valve handles; de-energizing motorized valves and mechanically restraining the valves.

Tags shall be attached to all isolation devices to give a visual indication that it is a means of isolation. The tags shall contain information on the purpose of the isolation, date of isolation, and details of the installer.

Draining, Venting, Purging and Flushing

As stated above, after the initial (or primary isolation) an equipment is required to be depressurized, drained and purged (if containing hazardous substance) before breaking containment to insert a blind or disconnect pipe spool. When final isolation has been achieved, the extent of further purging and flushing will depend on how hazardous the substance is and the type of intrusive maintenance activity to be carried out. If for example, the substance is flammable and the maintenance involves hot work activity, the equipment should be properly flushed with hot water (or steam) and in some cases with chemical additives (in line with the plant operations instruction manual). Gas tests will then be conducted to ensure that the atmospheric condition is safe to work, ensuring that representative samples are tested (e.g. top, middle and bottom portions of a vessel).

Testing and Monitoring Effectiveness of the isolation

It is important that the isolation is tested and proven to be effective before intrusive work is started. During the duration of the work, the isolation devices should be monitored. The isolation procedure should state method of proving each device, specifying the exact drain or vent or pressure indicator to use. Following this procedure, test the devices and confirm that they are holding (i.e. there is no leakage or pressure build-up inside the isolation envelope) before breaking containment. At the beginning of each shift and after every break, monitor the isolations by re-testing them using the same procedure before work commences.

Reinstatement of the plant

When maintenance work is completed, an inspection shall be conducted to ensure that all personnel, tools, equipment, maintenance debris, disused parts and other wastes have been removed. The reinstatement of the plant shall start by de-isolation in the reverse order in which the isolation was done. Before removing blinds or spades, it is important to test that the upstream isolation valve is still holding and there is no pressure build-up between the valve and blind. If this happens, a risk assessment shall be conducted and measures implemented to ensure that personnel are not exposed to hazardous substances before continuing with the de-isolation.

CHAPTER 17

MANAGEMENT OF CHANGE

Many major accidents in the petroleum industry over the past few decades have been traceable to a management of change (MOC) system that was either non-existent or badly managed. Change is sometimes desirable to make improvements, replace damaged parts or respond to an emergency. Management of change system ensures that careful consideration is given to safety, health and environmental implications that may result from a change. It is noteworthy that replacement with exact replica (replacement in kind) does not need to go through management of change.

Every petroleum process facility or organization should have a management of change procedure. The purpose of this procedure is to ensure that changes (except for replacement in kind) are identified, documented, properly reviewed, and approved by competent personnel prior to being implemented in order to avoid process accidents. It should ensure that consideration is given to the following before any change:

- Technical basis for the proposed change;
- Impact of the change on safety, health and environment;
- Change to operating procedures;
- Time frame for the change;
- Approval requirements for the proposed change.

It should be applicable to changes to process chemicals, equipment, facilities, technology and organization.

An MOC procedure should stipulate how to address different types of changes (emergency, temporally or permanent). A change that is initiated and implemented quickly for safety, health, environmental or potential equipment damage considerations when an emergency response plan has been implemented is known as emergency MOC. Temporary changes are those that typically should last for less than one year. Any change that is estimated to last more than a year should be considered as permanent.

The key steps of a management of change system are:

- Identify a change (if change is "replacement in kind", MOC is not required);
- Initiate MOC (set up a team in line with the procedure);
- Classify type of change (temporally, emergency or permanent);
- Assess or evaluate the impact (process hazard analysis);
- Identify action items (with responsibilities and timeframes);
- Approve change;
- Implement change (execute it, communicate, conduct pre-startup review);
- Close MOC;
- Conduct regular reviews.

It is important that employees that operate or maintain a facility whose job tasks may be affected by a change are informed and trained in the change before startup of the affected section of the facility.

In situations where a change affects the process safety information, such information is required to be updated accordingly. If the change requires modification of the operating procedures or practices, they should be updated as well.

CHAPTER 18

PRE-STARTUP SAFETY REVIEW

In the petroleum industry, major accidents have occurred during startups and shutdowns. In a bid to minimize this happening, pre-startup safety review (PSSR) should be carried out prior to introduction of feedstock or energy, after a change done in line with a management of change procedure, construction and turnaround/maintenance activities.

The PSSR is designed to ensure that all critical areas have been adequately assessed and addressed such that operation can start smoothly and safely. During the review, every part of the facility should be checked to be in a safe position and actions items arising from management of change process have been addressed.

The review should be conducted by a team of experienced personnel from various disciplines (e.g. engineering, operations, maintenance, safety, inspections) and they should check that:

- Construction has been completed in line with design specifications;
- Construction/maintenance wastes have been cleared and good housekeeping in place;
- Process hazard analysis was completed and recommendations addressed;
- Safety critical elements have been satisfactorily tested;
- Emergency response equipment are in place and adequate;
- Appropriate safety signs are in place;
- Actions arising from MOC (if applicable) have been addressed;

- All required process safety information is available and updated;
- Operating procedures (including startup) are in place and updated;
- Training of personnel on new or changed processes have been completed;
- Plant/equipment isolation is still in place;
- Plant/equipment has been cleaned or purged.

After the review, recommendations to address identified deficiencies should be categorized into pre-startup and post-startup actions. Pre-startup actions must be addressed before the equipment is put into operation. Post-startup actions can be done after operation is started, implying that they are not critical for safe startup. Before startup, a personnel with adequate competence and authority (depending on the complexity of equipment or plant) should review PSSR report, actions taken and give approval for startup to proceed.